Inbreeding and Outbreeding
The Genetic and Sociological Significance

by Edward M. East and Donald F. Jones

with an introduction by Jackson Chambers

This work contains material that was originally published in 1919.

This publication is within the Public Domain.

This edition is reprinted for educational purposes
and in accordance with all applicable Federal Laws.

Introduction Copyright 2016 by Jackson Chambers

Self Reliance Books

Get more historic titles on animal and stock breeding, gardening and old fashioned skills by visiting us at:

http://selfreliancebooks.blogspot.com/

Introduction

I am pleased to present yet another title on the Principles of Animal Breeding.

This volume is entitled "Inbreeding and Outbreeding" and was published in 1919.

The work is in the Public Domain and is re-printed here in accordance with Federal Laws.

As with all reprinted books of this age that are intended to perfectly reproduce the original edition, considerable pains and effort had to be undertaken to correct fading and sometimes outright damage to existing proofs of this title. At times, this task is quite monumental, requiring an almost total "rebuilding" of some pages from digital proofs of multiple copies. Despite this, imperfections still sometimes exist in the final proof and may detract from the visual appearance of the text.

I hope you enjoy reading this book as much as I enjoyed making it available to readers again.

Jackson Chambers

EDITORS' ANNOUNCEMENT

The rapidly increasing specialization makes it impossible for one author to cover satisfactorily the whole field of modern Biology. This situation, which exists in all the sciences, has induced English authors to issue series of monographs in Biochemistry, Physiology, and Physics. A number of American biologists have decided to provide the same opportunity for the study of Experimental Biology.

Biology, which not long ago was purely descriptive and speculative, has begun to adopt the methods of the exact sciences, recognizing that for permanent progress not only experiments are required but that the experiments should be of a quantitative character. It will be the purpose of this series of monographs to emphasize and further as much as possible this development of Biology.

Experimental Biology and General Physiology are one and the same science, by method as well as by contents, since both aim at explaining life from the physico-chemical constitution of living matter. The series of monographs on Experimental Biology will therefore include the field of traditional General Physiology.

JACQUES LOEB,
T. H. MORGAN,
W. J. V. OSTERHOUT.

PREFACE

IT is inevitable that each work planned as a member of a series of biological monographs should be somewhat technical. Of necessity each must be concise. In view of the difficulties these limitations involve, one may hardly expect to escape the criticism that the subject matter often tends to be esoteric in its nature, for few can say in Shaw's odd fancy, "I tried to do too much—and did it." Nevertheless, there has been a serious effort to avoid a mere record of the development of a specific problem in Genetics as an aid to the general biologist. No one could have a professional interest in a subject of this kind without the desire that there be some practical application of the results to agriculture and to the many phases of sociology where a knowledge of the laws of heredity is a first requisite. Though such applications of the genetic conclusions are touched but lightly here, there is the hope that the non-biological worker interested in problems of human welfare will find some new thoughts and pertinent suggestions in the compelling logic of the controlled experiments described throughout the pages. At least it was with this idea in mind that the authors prepared the first four chapters. For the zoölogist and botanist the well-known facts and elementary principles there discussed would have been unnecessary.

The manuscript has been the product, as it purports to be, of a very intimate collaboration, and the authors join in acknowledging their indebtedness to their fellow biologists for the privilege of copying several illustrations as noted in the legends, to Professor T. H. Morgan for helpful and suggestive criticism, and to Mr. L. C. Dunn and Mr. E. S. Anderson for assistance on the proofs.

E. M. E.
D. F. J.

BOSTON, September, 1919

CONTENTS

CHAPTER	PAGE
I. Introduction	13
II. Reproduction Among Animals and Plants	20
III. The Mechanism of Reproduction	36
IV. The Mechanism of Heredity	50
V. Mathematical Considerations of Inbreeding	80
VI. Inbreeding Experiments with Animals and Plants	100
VII. Hybrid Vigor or Heterosis	141
VIII. Conceptions as to the Cause of Hybrid Vigor	164
IX. Sterility and Its Relation to Inbreeding and Cross-breeding	188
X. The Rôle of Inbreeding and Outbreeding in Evolution	195
XI. The Value of Inbreeding and Outbreeding in Plant and Animal Improvement	210
XII. Inbreeding and Outbreeding in Man: Their Effect on the Individual	226
XIII. The Intermingling of Races and National Stamina	245
Literature	266

ILLUSTRATIONS

FIG.		PAGE
1.	Asexual Reproduction, an Amœba in Division	21
2.	Asexual Reproduction by Means of Runners	22
3.	Hermaphroditism in the Tapeworm Proglottid	24
4.	Rhopalura, an Example of Extreme Sexual Dimorphism	26
5.	Sexual Reproduction in Fucus	26
6.	Ulothrix, a Primitive Type of Sexual Reproduction	28
7.	An Adaptation for Self-pollination	30
8.	An Adaptation for Cross-pollination	34
9.	Diagram of Gametogenesis	38
10.	Diagram to Illustrate Fertilization	39
11.	Formation of Pollen Grains in the Lily	40
12.	Fertilization in the Embryo Sac of the Lily	41
13.	Entrance of Spermatozoön through Membrane of Egg of Star-fish	42
14.	Diagram Showing the Distribution of Sex Chromosome in Protenor	43
15.	Identical Quadruplets in Nine-banded Armadillo	44
16.	Diagram to Illustrate Inheritance of Sex-linked Character	48
17.	Diagram Showing Union of like Gametes	52
18.	Diagram to Illustrate Mendelism in a Cross between Long-Spiked and Short-Spiked Wheat	54
19.	Diagram to Illustrate Gamete Formation in a Dihybrid in Independent Inheritance	59
20.	Diagram to Illustrate Gamete Formation in a Dihybrid in Linked Inheritance	63
21.	Diagram to Illustrate Crossing-over	64
22.	Curves Showing the Limiting Values of the Coefficients of Inbreeding with Various Systems of Matings	84
23.	Graphs Showing the Total Inbreeding and Relationship Curves for the Jersey Bull, King Melia Rioter 14th	86
24.	Graphs Showing the Reduction of Heterozygous Individuals and of Heterozygous Allelomorphic Pairs in Successive Generations of Self-fertilization	90
25.	Graphs Showing the Increase in the Body Weight with Age for the Males of Inbred Albino Rats	107
26.	Graphs Showing the Increase in Weight of Body with Age for Different Series of Male Albino Rats	108

27. Graph Showing the Average Size of Litters Produced in Successive Generations of Inbreeding Albino Rats by Brother and Sister Matings.. 109
28. Goliath, an Albino Rat, the Product of Six Generations of the Closest Possible Inbreeding.. 110
29. Representative Samples of Inbred Strains of Maize after Eleven Generations of Self-fertilization............................. 130
30. Graphs Showing the Reduction of Variability and Segregation of Ear Row Number in Selfed Strains of Maize.................. 132
31. Plants of Maize after Eleven Generations of Self-fertilization and Their F_1 Hybrid.. 150
32. Ears of Maize after Six Generations of Self-fertilization and Their F_1 Hybrid.. 150
33. Graphs Showing Growth Curves of Two Inbred Strains of Maize and Their First and Second Generation Hybrids.................. 152
34. James River Walnut, Hybrid Between Persian Walnut and Butternut 154
35. Growth Curves of Parent Races and F_1 and F_2 Hybrids of Guinea Pigs 160
36. Diagram to Show How Factors Contributed by Each Parent May Enable the First Generation of a Cross to Obtain a Greater Development than Either Parent............................. 175
37. Cattaloes, the Product of Crossing the Cow and the Bison........ 180
38. Sterile Hybrid Between Radish and Cabbage..................... 192
39. Tassels of an Almost Sterile Strain Obtained by Inbreeding Maize.. 196
40. Representative Ears of a Cross Between Two Inbred Strains of Maize 202
41. Plants of a Cross Between Two Inbred Strains of Maize........... 202
42. Diagram Showing a Method of Double Crossing Maize to Secure Maximum Yields, Illustrated by Actual Field Results......... 203
43. First Generation Cross of Shropshire by Delaine Merino........... 212
44. First Generation Cross of Hereford by Shorthorn................. 216
45. "Big Jim" the Product of a Pure Bred Percheron Stallion Mated with a Grade Mare of the Same Breed....................... 220
46. First Generation Cross of Chester White and Poland China........ 224

INBREEDING AND OUTBREEDING

CHAPTER I

INTRODUCTION

INTEREST in the effects of inbreeding and of outbreeding is not confined to the professional biologist. Historically these are old, old problems, practical problems of considerable significance bound up with man's gravest affairs, his marriage customs and his means of subsistence. In these matters, moreover, the passing of time has not diminished the value to be attached to their solution. The questions involved belong to theoretical biology, it is true, and the professional biologist may lay claim to the first satisfactory analyses; but relatively his interest is that of yesterday, stimulated by the work of Darwin in establishing the doctrine of Evolution.

The intimate relation which the effects of various systems of mating bear to these three subjects will be seen more clearly from the following brief explanation.

Anthropological investigations have shown that many primitive peoples established rigid customs of exogamy—marriage outside the family or the clan. Such practices, after their identification with totemic systems by MacLennan, became the subject of much notable speculation. In particular may be mentioned the works of Frazer, Lang and Freud. Yet these writers have thrown little light on the origin of outbreeding as a social habit, and have contributed nothing whatever toward the solution of the questions of inbreeding and outbreeding in the sense in which they will be treated here. It is probable, indeed, that these customs usually originated without regard to matters of physical inheritance. The tribes concerned had seldom risen to a state of culture where the welfare of their descendants might be expected to cause anxiety, since in few cases had there been that development of animal husbandry necessary for the first glimpse into the mysteries of heredity.

These observations do not necessarily apply to the marriage folkways which developed in western Asia and Europe and were passed on to the United States. Our laws preventing marriages between certain degrees of kinship have been moulded by the touch of various civilizations, but in the main they are a legal heritage from the code of Hammurabi through the Hebraic Talmud. Since they are based largely upon the customs of pastoral nations, it may be they had *some* foundation in experience, half-truths drawn from casual and fragmentary observations of the shepherd and the cattleman. There is no historical record of such rational basis, however. Many of the conventionalisms rigidly stabilized by the hand of religious authority have not the slightest biological justification. Witness the English laws preventing marriage with a deceased wife's sister. On the other hand, if there had not been a dim but real fear of evil consequences arising from inbreeding, there would be something extraordinary in the frequencies with which taboos against consanguineous matings have persisted. Among the peoples contributing to European civilization, caste systems have been common, and the logical outcome of a caste system is marriage between near relatives. Pride of race encourages inbreeding among the ruling class, and power within that ruling class prompts the perpetuation of a serving class in the same manner. Why, then, should exogamy have been continued so commonly throughout epochs marked by rational thought and a high degree of culture? It is true, there are exceptions to this general rule. Rather intense inbreeding was practiced both in Egypt and in Greece when they were at the height of their power and influence. Nevertheless, exogamic customs have prevailed. They exist in Europe and America at the present day, and it is natural to wish to know whether there is any biological justification for them.

Let us propose three questions which will show the sociological bearing of the problems under consideration.

1. Do marriages between near relatives, wholly by reason of their consanguinity, regardless of the inheritance received, affect the offspring adversely?

2. Are consanguineous marriages harmful through the operation of the laws of heredity?

3. Are hereditary differences in the human race transmitted in such a manner as to make matings between markedly different peoples desirable or undesirable, either from the standpoint of the civic worth of the individual or of the stamina of the population as a whole?

Correct answers to these questions are a matter of more importance than a superficial consideration indicates. Settled in accordance with the biological facts, they aid in establishing a concrete scientific basis for marriage, divorce and immigration laws; they give grounds for predicting the changes to be expected in the body politic due to differential fecundity, birth control, and other agencies by which the character of the population is shifted; they even have some relevancy to many problems which one might suppose were wholly of an economic nature, such as minimum wages and mothers' pensions.

The second series of phenomena arousing interest in the results of inbreeding and outbreeding comes from observation upon domestic animals and cultivated plants. Plants are included by courtesy, though in reality intelligent plant breeding hardly began until the nineteenth century, and the methods adopted were taken from the procedures in use by animal breeders, with such modifications and improvements as the peculiarities inherent in vegetative propagation made necessary. Animal breeding, on the other hand, is a very ancient occupation, and more or less accurate data on the effects of interbreeding near relatives as compared with the effects of crossing different strains must have been collected by all of the old agricultural peoples. Since there is no question that under certain circumstances inbreeding does produce undesirable results—defectives, dwarf-forms, sterile individuals, etc.—it may be that their experience was at the base of *some* of the antagonism toward close-mating in the human race. Or, it is possible that early breeders observed the phenomenon, common to both animals and plants, that when two unrelated stocks are crossed the hybrids thus produced are often more vigorous than either parent—the phenomenon of hybrid vigor or heterosis, as it is called at the present time. There is no proof of such a sequence of ideas, but it seems to be a logical hypothesis. At any rate, the views of the animal raisers regarding

INTRODUCTION

inbreeding and the traditions regarding marriage of near kin are very similar. The great majority of breeders have an ineradicable fear of evil consequences if their matings are too close. Only here and there a few fearless ones have used systems of extremely close mating to perpetuate their breeds, and by such methods have built up invaluable races of horses, cattle, swine and poultry. But here a dilemma appears. Inbreeding has deplorable results in certain cases, yet in other instances the returns have been gratifying. What is to be the future practice? To be more than mere trial and error, it must be founded upon a cogent analysis of the whole subject.

Finally, interest in the effect of various systems of mating as natural phenomena has been stimulated by the study of organic evolution. The circumstantial data of comparative morphology show that in nature problems similar to those of man have arisen. If these problems are investigated some light may be thrown upon his difficulties. Sexual reproduction has been the most successful method of providing for the propagation of animals and plants. Does sexual reproduction, therefore, possess an advantage over other methods? Would it otherwise have persisted as it has in both kingdoms? It probably was not the original method of reproduction. Asexual reproduction, reproduction by simple vegetative division, appears to have held the stage when animals and plants were simple and unspecialized. Then, in all probability, came sexual reproduction with separation of the sexes. Secondarily, however, numerous species arose in both kingdoms wherein the sexes are united, male and female cells being produced in the same individual. Thus a system of mating entailing the greatest possible amount of inbreeding was established. But this system appears to have been deficient. Some evolutionary advantage associated with separation of the sexes was lost. There is reason for assuming that this advantage was connected with cross-fertilization, for tertiary developments in each kingdom brought about numerous mechanisms whereby cross-fertilization was established in hermaphroditic organisms. Still, in spite of the obvious success of bisexual and of cross-fertilized species, as shown by their frequency, numerous self-fertilized species, and even species reproducing exclusively by asexual methods have kept their places in the struggle for existence. Both inbreeding and outbreeding systems have developed side-by-side under natural conditions. The data of comparative morphology, therefore, seem as contradictory as those from anthropology and agriculture.

The puzzles presented by these general facts taken from history, husbandry and biology have one common feature. They centre on the problem of inheritance. Fortunately, though less than two decades have passed since the application of quantitative experimental methods to biology became somewhat general, the mechanism of heredity is no longer a riddle; and to-day the effect of inbreeding and outbreeding on plants and animals can be described in considerable detail and interpreted with singular precision. Having this interpretation it may be applied to the three fields of interest we have described.

In the ensuing pages the important controlled experiments in inbreeding and outbreeding necessary for an orderly and consistent interpretation of the facts are discussed. Uncontrolled experiments, casual observations of stock breeders, data on human marriages between near relatives, have been omitted designedly. Numerous data of these types have been available for many years, but they have been of little service in clarifying the situation. This is not altogether due to their fragmentary character in point of time, or even to the fact that they usually lack the precision necessary in data to be used in the analysis of such complex phenomena. Data for a limited number of generations are often useful, and precision is a relative matter. The truth is, the majority of these records was collected without regard to the type of fact required, and without reduction to concrete numerical terms. In other words, in records otherwise accurate, critical data are omitted; and those given are relatively useless on account of their form.

A detailed application of our conclusions to sociology, agriculture and evolutionary theory has not been attempted. It is hoped that the suggestions along these various lines are sufficient to show how such application can be made; but human direction of evolution either in man or in the lower organisms is beset with difficulties so numerous and so prodigious that each problem must have its individual solution.

CHAPTER II

REPRODUCTION AMONG ANIMALS AND PLANTS

In order to obtain a proper orientation of the problem of inbreeding and outbreeding, one must consider first some of the general facts regarding reproduction among animals and plants and their relation to inheritance.

The significant changes in both kingdoms have been remarkably similar. The differences are differences in detail, and for this reason they are additional arguments in favor of the idea that there are special advantages associated with the coincidences found in the general processes involved. For example, asexual propagation is more general in the simpler, sexual reproduction in the higher organisms. But sexual reproduction in animals has largely supplanted the asexual method, in plants sexual reproduction was merely added. Is this not evidence of an importance to be attached to the sexual method, apart from a simple provision for multiplication? Again, the diversity of sex organs which has arisen among the various groups of animals and plants is highly surprising, yet this dissimilarity may be wholly of a superficial nature. When examined solely with the object of inquiring what systems of mating these variations entail, the parallelisms in each history stand out impressively. If these facts be kept in mind throughout the short discussion of heredity and reproduction which follow, their probable evolutionary significance is not difficult to grasp; but if the profusion of variation in detail, or even the general mechanism of accomplishing a particular result is allowed to distract attention, the *end* may be lost to sight through admiration of the ingenuity of the *means*.

There seems to be no question but that sexual reproduction is a more recent means of propagation than asexual reproduction. Although asexual reproduction in the narrow sense, that is, by means of simple division or by budding, is common among the protozoa, the sponges, the cœlenterates and the flat worms, it becomes sporadic in the molluscoids and annelids, and is found in only one or two isolated instances in forms as highly specialized as the arthropods and the chordates. If fragmentation succeeded by regeneration of the lost parts be conceded to be a true means of reproduction, however, echinoderms and nematode worms are included. Thus of all the great groups of animals only certain worms (*Trochelminthes*) and the molluscs have no asexual reproduction in the usual sense of the word, and zoölogists would

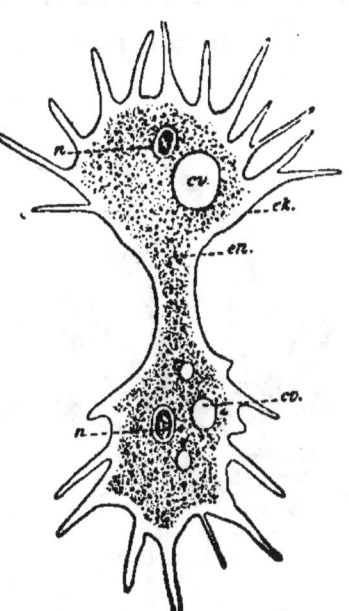

Fig. 1.—Asexual reproduction. An amœba in division. *cv.* contractile vacuole; *ek*, ectosarc; *en*, entosarc; *n*, nucleus. (Kingsley after Schulze. Courtesy Henry Holt & Co.).

hardly feel safe in maintaining its absence in these two phyla since the life history of so many forms is unknown. But since asexual reproduction is replaced by sexual reproduction to a greater and greater extent as the higher forms are reached one cannot avoid the conclusion that the latter has proved to be the really successful means of propagation. Nevertheless, variations appeared in highly specialized forms which permitted return to an asexual type of reproduction. In the arthropods, as well as in some other forms, mechanisms arose by which the eggs developed without fertilization. This parthenogenetic reproduction has been relatively successful, but only as a stop-gap. Sexual reproduction persists and is used as an occasional means of propagation. It would seem that it possessed advantages too great to be given up entirely.

Even as sexual reproduction is a later method of propagation than asexual reproduction, hermaphroditism appears to be a secondary development from forms in which the sexes were separate (gonochorism or diœcism). Omitting the protozoa in which it is difficult to decide such sexual differences, gonochorism is present in every great animal group but the sponges, and hermaphroditism everywhere except in the *Trochelminthes*, although in *Nemathelminthes*, *Echinodermata* and *Arthropoda* it is rare. An extended experiment on the subject of hermaphroditism certainly was made, but that it was an experiment, that hermaphroditism is from the evolutionary standpoint a secondary institution, is clear if one considers the anatomical evidence, as is shown by Caullery.[23] Generally, hermaphroditism is a condition associated either with parasitism or with a sedentary life. Furthermore, hermaphroditic organisms do not have a truly simple organization. They have a superficial simplicity, due to an adaptation to their mode of life, but if one compares hermaphroditic and gonochoristic species group

Fig. 2.—Asexual reproduction by means of runners in the hawkweed. (After Andrews.)

by group, for example unisexual land or fresh-water worms with their bisexual marine cousins, he finds the former to be the more complex, particularly as to their sex organs. The fact that the sponges are hermaphroditic might be considered as weighing against this argument, but it is not without the bounds of probability that the sponges are further along in specialization than is generally admitted, for to find the substance nearest chemically to the so-called skeleton of the sponges, one must search among the arthropods—the product of the spinning glands of certain spiders and insects.

Hermaphroditism, pure and simple, however, was not a success. Only a few degenerate forms retained self-fertilization and persisted. Among them may be mentioned the tapeworms, certain crustaceans (*Sacculina*) parasitic on crabs, and the colonial forms, bryozoans and tunicates, the latter being perhaps the most degenerate of all animals since they are wholly unrecognizable as relatives of the vertebrates except at one short stage of their life history. In most of the hermaphroditic types new characteristics appeared which enabled them to exercise one of the important functions of bisexuality, cross-fertilization, without giving up the obvious energy conservation attainable through the production of both sex cells in a single individual.

In nearly all of these forms, this was made possible by the development of the eggs and of the sperm at different times. In a few isolated cases among the turbellarians and the tunicates the eggs developed first and then the sperm; the animal is first a female and later a male (protogyny). But in a greater number of species, the individual is first a male and afterwards a female (protandry). In the tapeworm (Fig. 3), for example, each segment contains a complete reproductive system, testes, ovaries and accessory glands; when young the testes function, when older the testes atrophy and the ovaries develop. In some of these protandrous species there is even a change in the whole structure of the body, includ-

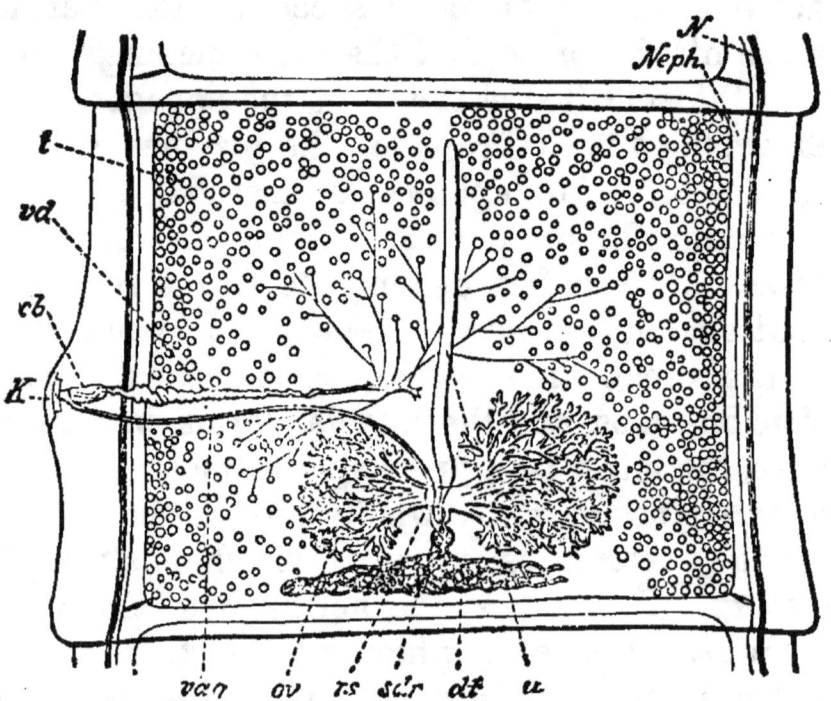

Fig. 3.—Hermaphroditism in the tapeworm proglottid. *K*, genital pore; *ov*, ovary; *rs*, receptaculum seminalis; *t*, testes; *u*, uterus; *vd*, vas deferens. (Kingsley after Sommer).

ing the sexual orifices. The isopods of the family *Cymothoidæ*, a group of crustaceans parasitic on fish, furnish a beautiful illustration. In the male stage the animal is a typical crustacean and would be recognized as such by any layman with a very slight knowledge of zoölogy; but when the animal passes over into the female stage it becomes merely a great egg sac many times the previous size. One would hardly suppose the two stages belonged to the same order, not to mention a transformation of the same individual.

A few other mechanisms which promote cross-fertilization have been found in isolated cases. They are not as widespread as the one just described, but are peculiarly interesting nevertheless. Among certain of the cirripedes, the normal individuals are hermaphroditic, but in addition a few tiny degenerate males are developed. They are little more than bags of sperm and are calculated to make somewhat amusing any generalization as to the "stronger sex." Darwin, who discovered them, called them complemental males. Another means of preventing continued self-fertilization is self-sterility, a condition in which self-fertilization is very difficult or even impossible through some physiological impediment which is not clearly understood. It was demonstrated by Castle [17] for the American race of *Ciona intestinalis*.

In what appear to be the essential features, the vicissitudes of reproduction have been similar in the vegetable kingdom. The problems were solved in different ways, but the gross results are largely the same. The most striking difference is the varied success of certain mechanisms. In the animal kingdom sexual reproduction wherever instituted practically always displaced asexual reproduction. Only in a few forms which are either fixed or parasitic in their mode of life did the two methods persist side by side. In plants, however, where the sessile is the common condition, asexual and sexual reproduction have continued harmoniously side by side clear up through the angiosperms. Again, there is a marked difference in the success of hermaphroditism. In plants hermaphroditic forms became the dominant types in the highest and most specialized group, the seed plants, while in the highest group of animals, the mammals, only an occasional individual showing rudimentary hermaphroditism is found.

Just when sexual reproduction first originated in the vegetable kingdom is even more of a question than among animals. Only a few very simple types, the schizophytes (bacteria) and myxomycetes, have passed it by. Perhaps

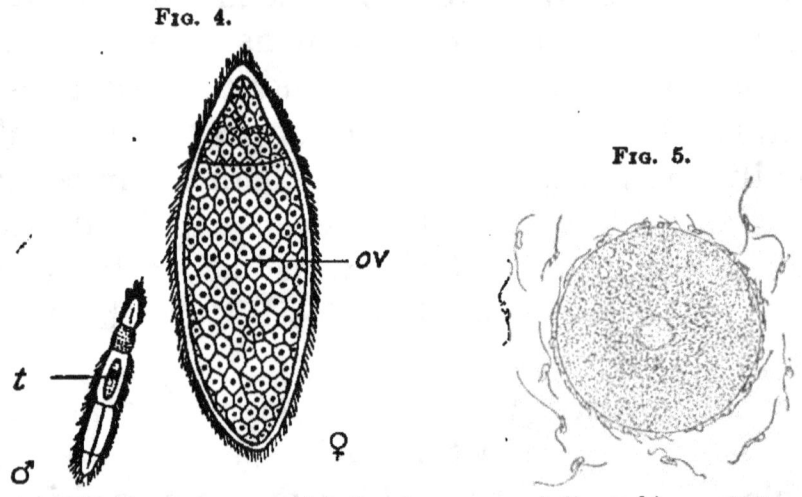

Fig. 4.—*Rhopalura*, an example of extreme sexual dimorphism. (After Caullery.)
Fig. 5.—Sexual reproduction in *Fucus*, giving some idea of the difference in size of egg and sperms. Sperms should be about one-tenth the size shown. (Bergen and Davis after Thuret.)

it is for this reason these forms have remained the submerged tenth of the plant world. It is tempting, as Coulter [32] says, to see the origin in the *Green Algæ*. There, in certain species, of which Ulothrix is an example (Fig. 6), spores of different sizes are produced. The large ones having four cilia are formed in pairs in each mother cell, the smaller ones usually having two cilia occur in groups of eight or sixteen in each spore-producing cell. Those largest in size germinate immediately under favorable conditions and produce new individuals.

Those of lesser size also germinate and produce new individuals, but these are small and their growth slow. Only the smallest are incapable of carrying on their vegetative functions. These come together in pairs and fuse. Two individuals become one as a prerequisite to renewed vigor. Vegetative spores become gametes. Something valuable—speed of multiplication—is given up that something more valuable in the general scheme of evolution may be attained.

This is indeed an alluring genesis of sex. It is rather *a* genesis of sex, however, than *the* genesis of sex. Various manifestations of sex are present in other widely separated groups of unicellular or simple filamentous plants, the *Peredineæ*, the *Conjugatæ* and the *Diatomeæ*—the *Conjugatæ* being indeed the only great group of plants in which there is no long continued asexual reproduction. In these forms one cannot make out such a good case of actual gametic origin, but the circumstantial evidence of sex development in parallel lines is witness of its paramount importance.

After the origin of sex, many changes in reproductive mechanisms occurred in plants, but most of them resulted merely in better protection for the gametes, in increased assurance of fertilization, in provision for better distribution, or in greater security for the young plant.

First, perhaps, there was physiological differentiation of the gametes. At least such an interpretation may be given to the form of conjugation found in *Spirogyra* and other *Conjugatæ*, where, either by solution of the wall separating them, or by the formation of a tube-like outgrowth of one or both cells so that the ends touch, the contents of one cell pass over to the other. We may think of the stationary cell as female and the other as male.

Another line of development, however, became the dominant one in the plant kingdom just as it did in the animal world. A morphological differentiation of the sex cells occurred. One became a large inactive cell stored

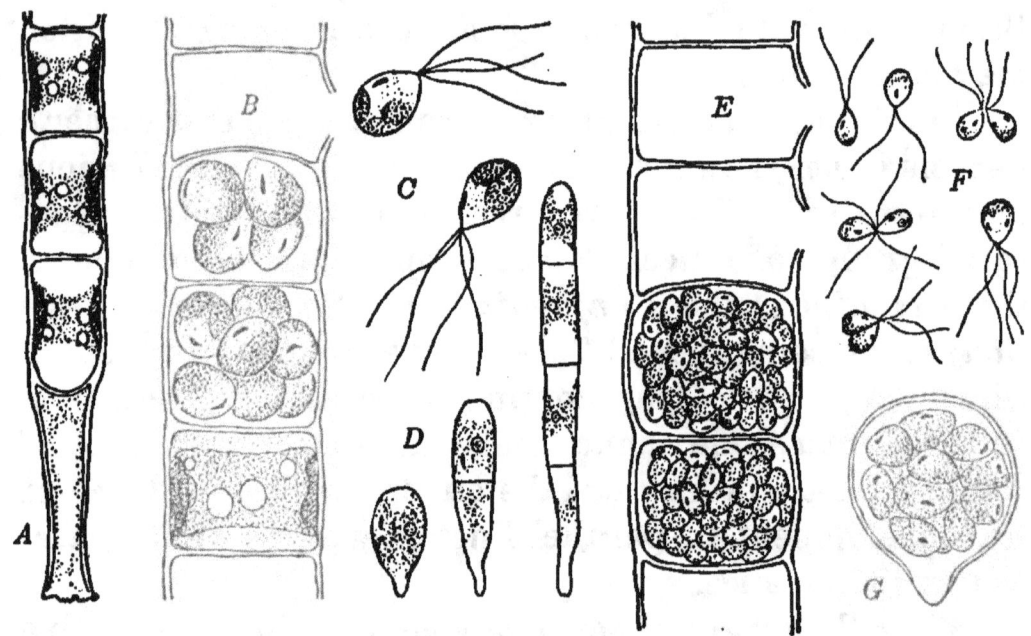

Fig. 6.—*Ulothrix*, a primitive type of sexual reproduction. *A, B*, filaments; *C*, zoospores; *D*, germination of zoospore; *E*, gamete formation in filament; *F*, gametes and their fusion; *G*, germination of zygospore. (Bergen and Davis after Dodel.)

with food, the egg; the other became small and motile, the sperm. This change is well illustrated in *Fucus* (Fig. 5), one of the brown algæ. It is clear that such a change increased the probability of fertilization, since many sperms could be produced without utilizing a great deal of energy, and since the attraction of the egg for the sperm was presumably augmented.

A further stage in the evolution of sex was reached when the cells producing the eggs or the sperms were differentiated, thus providing for protection of the gametes. Such organs of various types and known by different names have persisted throughout all the higher plants. One may call them ovaries and spermaries and thus keep in mind that in animals the same types of change occurred.

The final step in the general development of sexuality is the restriction of the formation of sex organs to a particular phase in the plant's life, which on this account is known as the gametophyte. The remaining stages are known as non-sexual or sporophytic, because they are characterized by the production of non-sexual reproductive cells, the spores. The liverworts, the mosses, the ferns and the seed plants are thus set apart.

Since these two phases alternate with each other, pairs of reproductive cells of the gametophyte producing the sporophyte, and the non-sexual spores of the latter giving rise to the gametophyte, the sequence has retained the name of *alternation of generations.*

In the higher liverworts and mosses the gametophyte carries on the greater part of the nutritive work, but in the ferns the sporophyte becomes the dominant structure; while in the seed plants the gametophyte has degenerated until it consists of but two or three cell divisions.

There is no question but that all of these numerous changes are merely insurance against the future, something that may be said of seed production as a whole, since the seed is but the younger generation nourished on the parent stem. And it is interesting to note that just as animals and plants paralleled each other in gamete protection and provisions for assuring fertilization, so also the final step in each, the mammals and the seed plants, was the protection of the young. In certain particulars, however, the higher plants did not simulate the higher animals in their reproductive evolution, and it is not difficult to see the reason for the divergencies. Plants retained asexual reproduction as an alternative method of propagation, and made a success of hermaphroditism. The obvious necessity for both was their fixed condition, their slavery to the soil; but if hermaphroditism with its simplest implication, self-fertilization, had become domi-

FIG. 7.—Adaptation for self-pollination by means of spiral twistings of stamens and style. (After Kerner.)

nant, there would have been little from their life histories upon which to base an argument regarding the respective virtues and defects of inbreeding and outbreeding. This, however, was not the case. Many plants characterized by autogamy persisted and flourished. They even developed numerous devices promoting self-fertilization (Fig. 7), such as pollination before the flower opens, inclination of the anthers toward the pistil or the pistil toward the anthers, rapid elongation of the pistil through a ring of stamens, or various torsions of the accessory floral parts; but it seems perfectly clear from the exhaustive investigations on the fecundation of plants made in recent years that only an extremely small percentage of the species of flowering plants which have held their own to the present day in the struggle for existence, have adopted a method of fertilization which permits no crossing. Some of our most vigorous cultivated plants—tobacco, wheat, peas and beans—are naturally and usually self-fertilized, but they each and every one have their flowers so arranged as to permit an occasional cross.

At the same time, one would be too hasty if he concluded from these facts that continuous self-fertilization or other means of reproduction which result in a single line of descent is incompatible with inherent racial vigor. At least, there is evidence that various species which seem well able to hold their own seldom resort to crossing as a means of propagation, yet one could hardly use them as examples of degeneration. As illustrations, there is no need to go below the flowering plants, either, although if one desires an example of a long-continued evolution of species and genera without any form of sexual reproduction he is forced to look to the Basidiomycetes. In this large group the fungi are not only asexual themselves, but appear to have been developed in a purely asexual manner from asexual ancestors. But in the flowering plants, many of our most useful types—the potato, the banana, hops and sugar cane—seldom have recourse to sexual reproduction. It is true many agriculturists insist that these species sooner or later degenerate for this very reason, but they have never been able to bring forward one atom of critical evidence to uphold their view. Varieties of potatoes or of sugar cane do indeed degenerate, but it is probably because of disease which from their method of propagation is difficult to eradicate, and not because of the method itself. Again, if one desires further evidence of descent in a single pure hereditary line consistent with high specialization and inherent vigor through long periods of time, there is the phenomenon of apomixis to be cited. Apomixis is a general term for certain reproductive anomalies in plants which are really a return to vegetative reproduction. In a broad way it is synonymous with parthenogenesis in animals; but parthenogenesis in animals includes only reproduction from an unfertilized egg, while apomixis takes in reproduction from certain cells which are not eggs. Some twenty or thirty species of vascular plants have already been found to reproduce in this manner, and unquestionably the list is very incomplete. Examples from *Polypodiaceæ*, *Ranunculaceæ* and *Rosaceæ* are not uncommon, but in particular it is the *Compositæ*, the highest group of flowering plants, which seem inclined to make this method of reproduction a habit. Of course, one cannot insist that such a return to primitive reproductive methods even by a more modern labor-saving route is wholly for the good of the species concerned. No one in possession of all of the facts could maintain the change to be progressive, or argue that the species adopting it will have a great future as future is measured by the evolutionist. This is not the contention. We merely cite the adoption of apomixis by flourishing genera of the most specialized and highly developed plants as examples of asexual reproduction over long periods without visibly harmful effects. We do this because we believe the emphasis put by Darwin and his followers on supposed ill effects following any type of inbreeding or asexual propagation was misplaced. Certainly the majority, the great majority, of the higher plants returned to a type of reproduction which held all the advantages of bisexuality by evolving means for promoting cross-fertilization. But it is the advantage of cross-fertilization and not the assumed disadvantage of self-fertilization that should be stressed. The Knight-Darwin Law, "Nature abhors perpetual self-fertilization," should read *Nature discovered a great advantage in an occasional cross-fertilization.*

The higher plants made a success of hermaphroditism because there was a return to the advantages of gonochorism through the development of almost innumerable devices tending to promote frequent crossing between plants of the same and nearly related species. Some species did actually return to true structural gonochorism, but in most cases other means of obtaining cross-fertilization were developed. There was no advantage, considering their sessile mode of life, in relinquishing the possibility of self-fertilization.

Some of the various cross-fertilization mechanisms utilized are very reminiscent of those of animals. Monœcism, the production of male and of female flowers on the same plant, and dichogamy, the maturation of the male and female organs at different times, have their counterparts in the other kingdom. So also the physiological phenomenon self-sterility of which only one instance is known among animals is very common among plants. Some hundred or so species distributed throughout thirty-five or more families have been shown to be self-sterile, although the true number is probably many times this figure. Again polygamy, where, in addition to hermaphroditic flowers, either male or female flowers are developed, has its analogue in the complemental males characteristic of the Cirripedes.

But by far the most numerous and most interesting adaptations for cross-pollination are characteristic of plants alone. These are the thousands of structural modifications which utilize external agencies. Wind and water have not been despised, but the real servants—they are

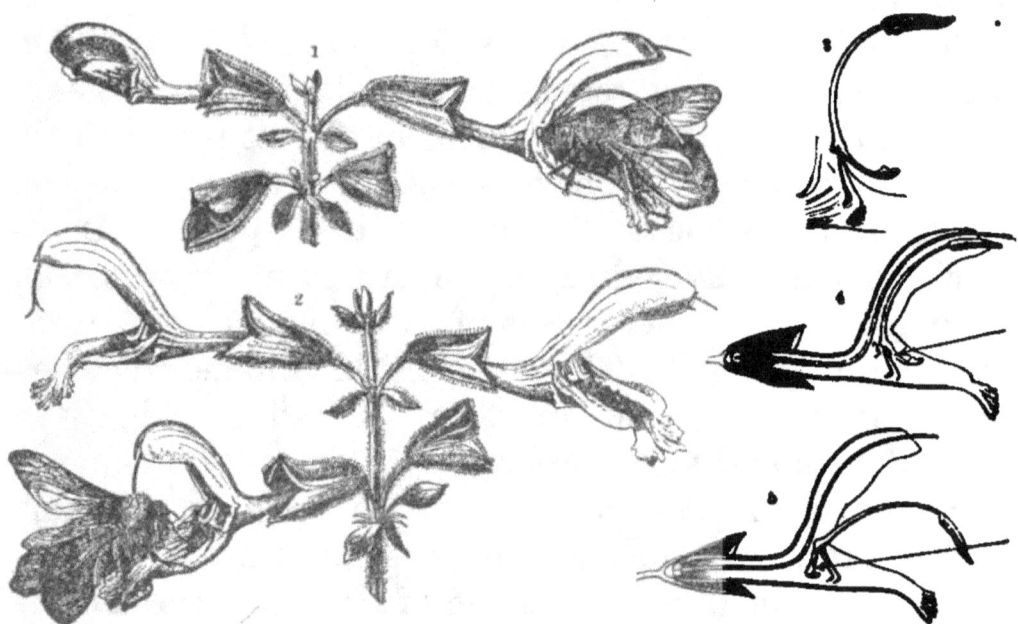

FIG. 8.—Adaptation for cross-pollination, transference of pollen by insects. (After Kerner.)

not slaves for they are paid for their services—are the lower animals and in particular the insects.

The ideas of Darwin resulting in the tremendous labors of Müller, Delpino, Kerner, Knuth and others have made it no longer necessary to describe the facts concerning the dispersal of pollen by animals.[124, 157] The subject has been so fascinating that it is common knowledge how the insects are attracted to flowers by odor and by color; how they are rewarded for visits by nectar and by pollen; how provisions are made to use them as pollen carriers through numberless modifications of calyx, corolla, stamens and pistil; how the animals themselves have developed organs for extraction of food or for attachment to the blossoms (Fig. 8). Perhaps some of these mutual adaptation mechanisms are a little fanciful, but the fact remains that actually an occasional or a frequent cross-pollination is secured by a majority of our 100,000 or more species of flowering plants by means of insects, and the hundreds of mechanisms by which it is obtained are witness of its paramount importance.

The thesis of this chapter, then, is simple. The whole trend of evolution in both animals and plants as regards all the mechanisms in any way connected with reproduction, has been such as to provide effectively for continuous descent. In the midst of strenuous competition for place, those organisms which were able to cross with others, at least occasionally, held such an advantage over those which were compelled to continue through one single line of descent, that their descendants have persisted in greater numbers. They have dominated the organic world. Any satisfactory interpretation of the effects of inbreeding and outbreeding must permit a reasonable explanation of this situation.

CHAPTER III

THE MECHANISM OF REPRODUCTION

There is a division of labor in all the higher plants and animals, the result of setting apart definite tissues for producing germ cells. In addition, another important matter is accomplished. The germ cells are insulated from ordinary environmental changes, and are enabled to go through a very exact routine of processes in preparation for the formation of the new organism—the zygote.

In general the animal body or the sporophyte of the higher plants can be considered as a double organization. Various parts make up each of the cell units; but of them all the *nucleus,* and within the nucleus the *chromosomes,* seem to be the most important. Each species has a characteristic and constant number of these bodies, and it is their distribution which parallels—and probably regulates—the distribution of the hereditary differences within a species. The double organization of the bodies of the higher organisms is dependent upon the receipt of *one* set of these chromosomes from *each* parent. And it is the peculiar method by which these chromosomes are apportioned to the gametes, together with experiments on the actual distribution of characters in the generations succeeding a cross, which have given us a fairly clear idea of heredity as a mechanical process.

In ordinary cell division during growth each chromosome divides longitudinally so that both daughter cells apparently receive an exact half of the chromatin, although possibly some sort of a special apportionment is made in the segregation of particular tissues. But when the germ cells are formed, at spermatogenesis and oögenesis, the chromosomes unite in pairs, a process technically known as *synapsis,* and at division one member of each pair passes entire to one of the two new daughter cells, thus reducing the number of the chromosomes in the gamete to one-half of those possessed by the body cells. Subsequently there is an *equating* or halving division similar in appearance to the cell divisions in ordinary growth. Four gametes are thus formed. Leaving out of account the behavior of certain chromosomes believed to control the distribution of sex, there is good evidence that this union of chromosome pairs at synapsis *always* takes place between two chromosomes, one of which had been received from the father and one from the mother. In other words, it seems clear that each gamete obtains *one of each kind of chromosome,* although it is a matter of chance whether the cell receives the maternal or paternal representative of any type. Thus, if the chromosomes of the body cells of a particular species are six in number, and we represent them as *ABC abc,* regarding *A, B,* and *C* as of maternal and *a, b,* and *c* as of paternal origin, at synapsis *A* only pairs with *a, B* with *b* and *C* with *c.* This procedure, however, will yield eight types of gametes, *ABC, ABc, AbC, aBC, Abc, aBc, abC,* and *abc,* since it is a mere matter of chance which daughter cell receives either member of any pair.

In spermatogenesis four sperms are formed from each immature germ cell, but in oögenesis—the maturation of the egg—only one functional gamete is produced, the other three being aborted. Nevertheless, the two processes are similar in all essential features, as may be seen in Fig. 9, the elimination of three out of four of the oöcytes taking

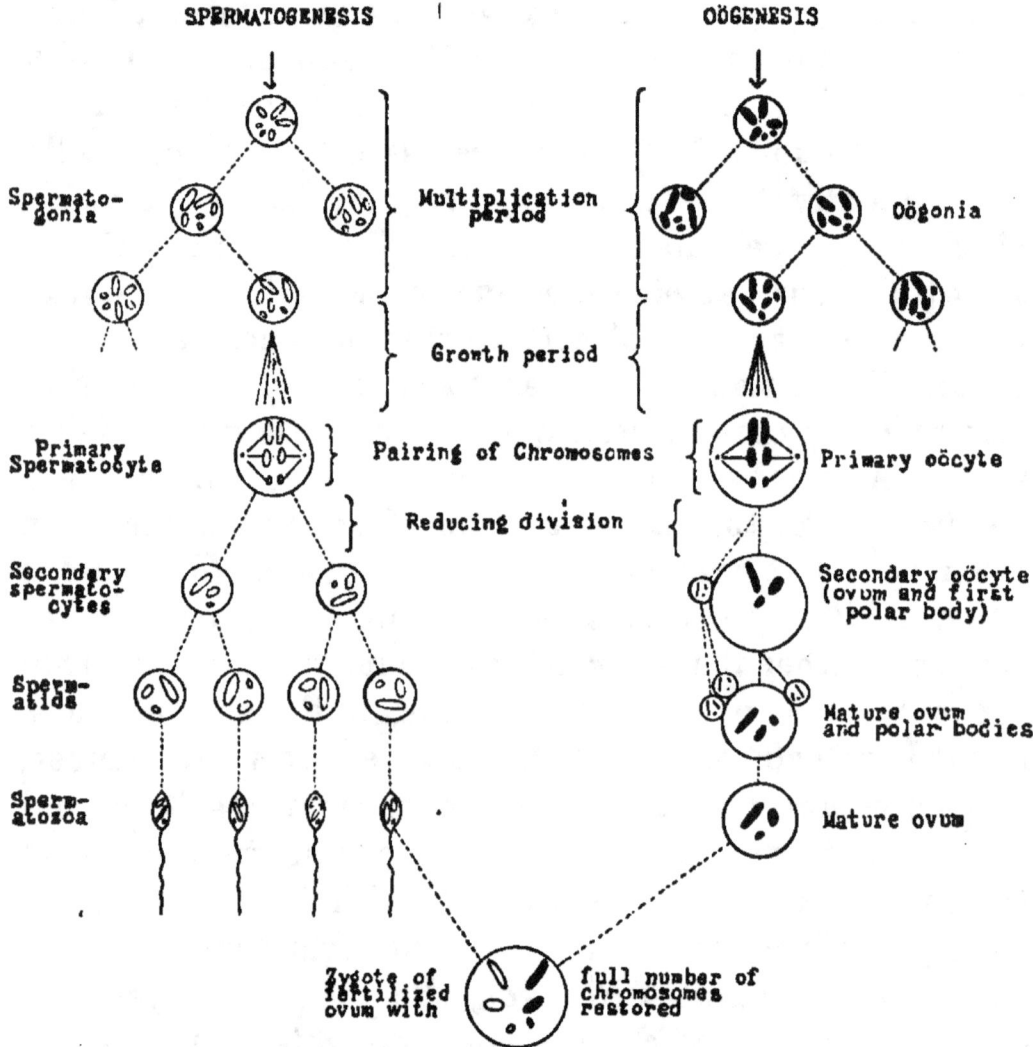

Fig. 9.—Diagram of gametogenesis showing the parallel between maturation of the sperm cell and maturation of the ovum. (After Guyer.)

place in order that their store of nutritive materials may go to make one large egg.

Fertilization consists in the fusion of one egg with one sperm, thus bringing back the double number of chromosomes characteristic of the body cells (Fig 10), and since it is a matter of chance what gametes unite, such gametic differences as we have illustrated would give a possibility of obtaining 8 × 8 or 64 types of zygotes.

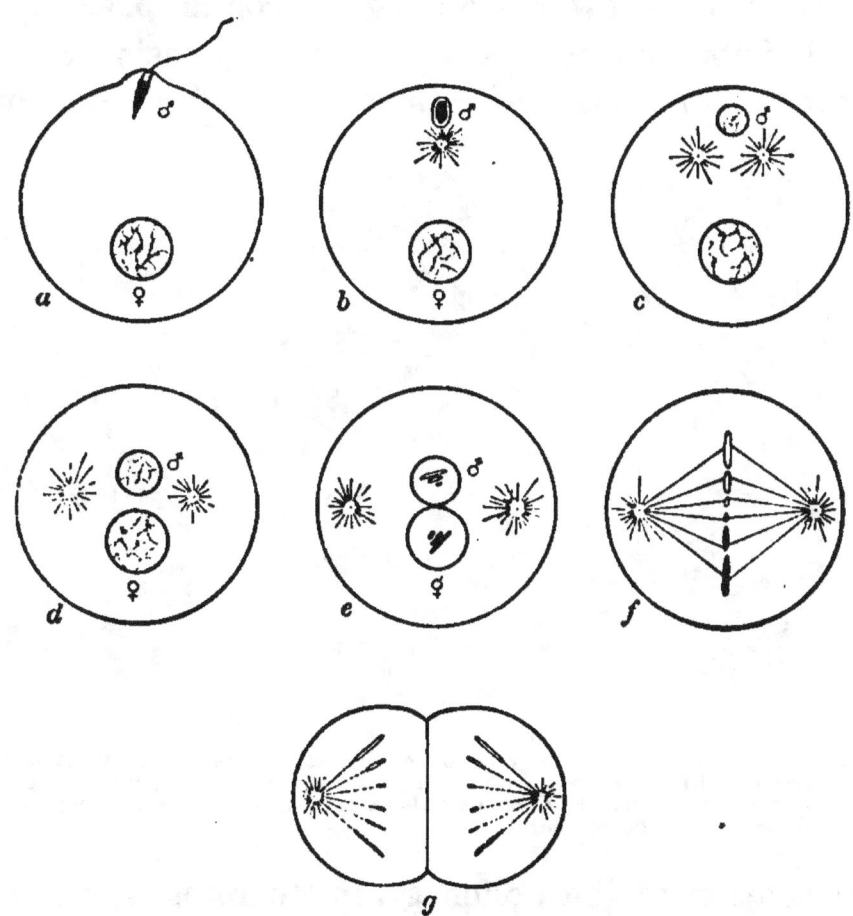

Fig. 10.—Diagram to illustrate fertilisation; ♂, male pronucleus; ♀, female pronucleus; observe that the chromosomes of maternal and paternal origin, respectively, do not fuse. (After Guyer.)

The mechanism of the process of gametogenesis and fertilization in animals need not concern us further here. We must speak of the process in the seed plants, however, for a rather odd phenomenon occurs there to which there will be occasion to refer later.

Reduction of the chromosomes takes place in plants just as it does in animals, but the introduction of a gamete generation, the gametophyte, complicates matters. In the seed plants, pollen mother cells are produced in the anthers of the flower which go through precisely the same divisions as in animal spermatogenesis (Fig. 11). But each of the four nuclei thus produced divides during

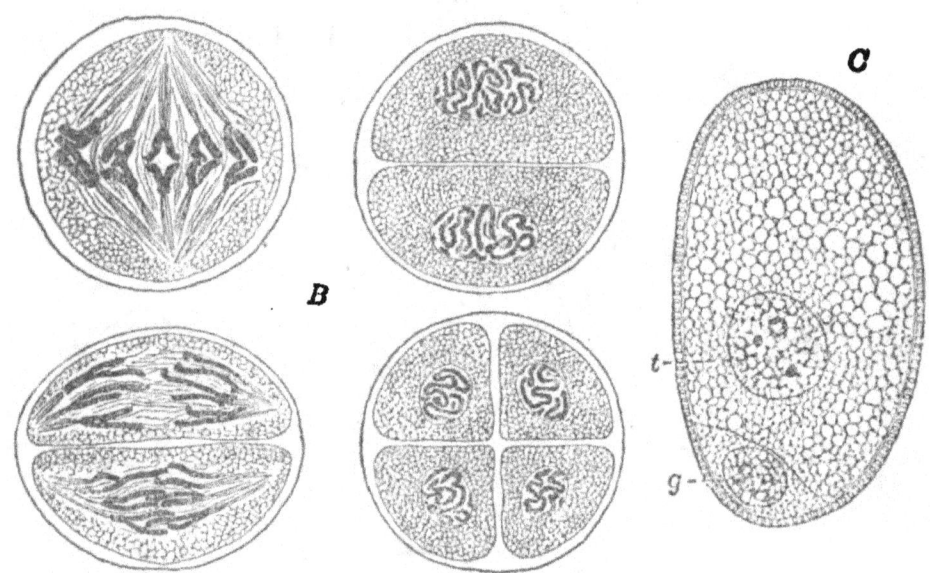

Fig. 11.—Formation of pollen grains in the lily. *B*, stages in the formation of pollen grains in a group of four (tetrad) within the pollen mother cell; *C*, mature pollen grain with early stages in the development of the male gametophyte; *t*, tube nucleus; *g*, generative nucleus. (After Bergen and Davis.)

the formation of the pollen grain, forming a *generative* and a *tube* nucleus. The tube nucleus it is that germinates and passes down the style when the pollen grain falls on a ripe stigma. During this period of pollen tube growth the generative nucleus passes down through it toward the ovule, and while so doing divides again, leaving two nuclei each with a function to perform. One fuses with the egg and the other with the so-called endosperm nucleus, completing in this manner the peculiar *double* fertilization characteristic of the angiosperms.

In the meantime, the *egg* and the *endosperm nucleus* have been prepared by the necessary cell divisions of the female gametophyte. The reduction division occurs in the usual manner, and as in animals three of the cells are absorbed, leaving a single one to provide for the hereditary succession. Its container enlarges and becomes the embryo sac, while the cell itself typically goes through three cell divisions resulting in the formation of eight nuclei. Any of these nuclei may become the egg, but generally the egg can be recognized by its position (Fig. 12). Two others from among these nuclei fuse together and become the endosperm nucleus, which in turn fuses with the second male nucleus and by succeeding cell divisions forms the endosperm of the seeds, the function of which is to furnish food for the young plant, the embryo. Thus, if we represent the chromosome complex of the gametes by x, the embryo is $2x$, and the endosperm $3x$.

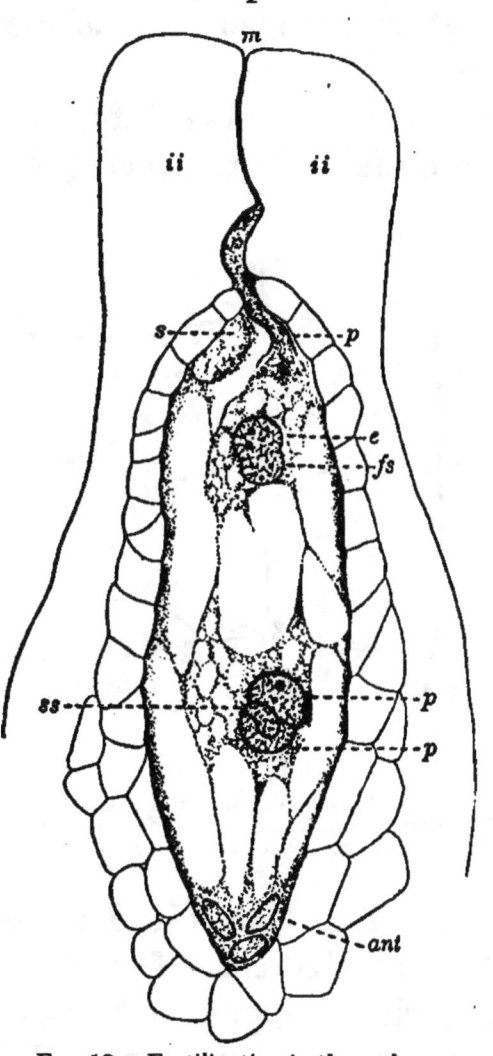

Fig. 12.—Fertilization in the embryo sac of the lily. *e*, egg; *fs*, first sperm; *pp*, fused polar nuclei; *ss*, second sperm. (After Bergen and Davis.)

It is clear from this short description of gametogenesis and fertilization that the processes in plants and in animals are identical in what we deem to be the essential features, the behavior of the chromosomes. If one visualizes the behavior of hereditary characters in crosses in which the parents differ as the result of the operation of potential factors carried by these bodies, he can correlate

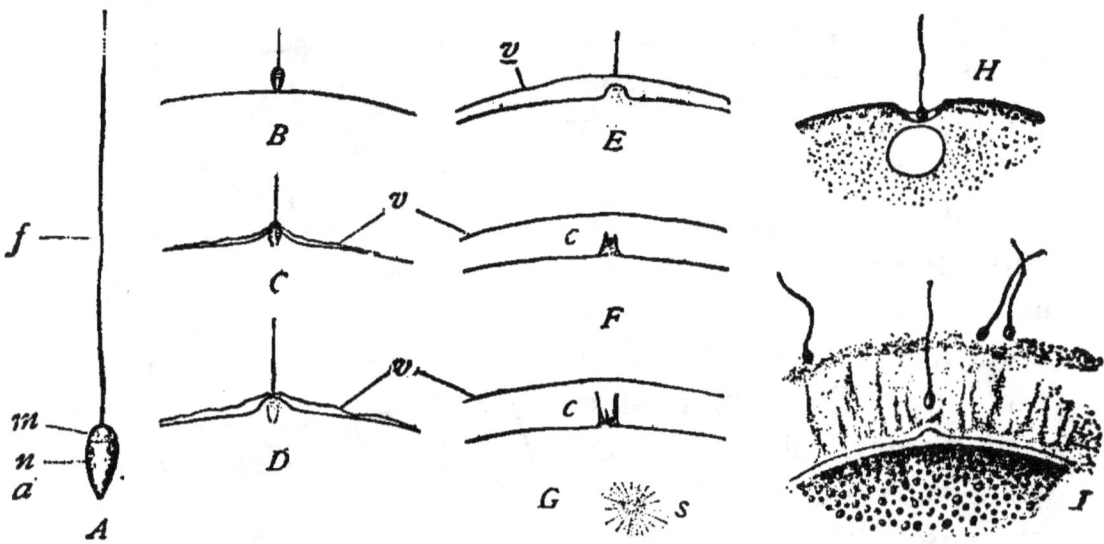

Fig. 13.—Entrance of the spermatozoön through the membrane of the egg of a starfish giving an idea of the difference in size between the sperm and the egg. (Wilson's "The Cell." Courtesy Macmillan Co.)

every fact thus far discovered, with the exception of a few isolated cases found in plants where particular characteristics appear to be distributed by the cytoplasm lying outside of the nucleus. Not only can the distribution of ordinary characters be interpreted as functions of the chromosomes, but the distribution of the sexes as well. There is reason to think the behavior of the sex-controlling chromosomes may perhaps occasionally be influenced by external conditions, but sex itself is determined by the

THE MECHANISM OF REPRODUCTION 43

behavior of particular chromosomes of which we have not hitherto spoken (Fig. 14).

The evidence in favor of this view of the determination of sex at the time of fertilization through the chromosome complex is from several very different sources.

First, there is the phenomenon of multiple births

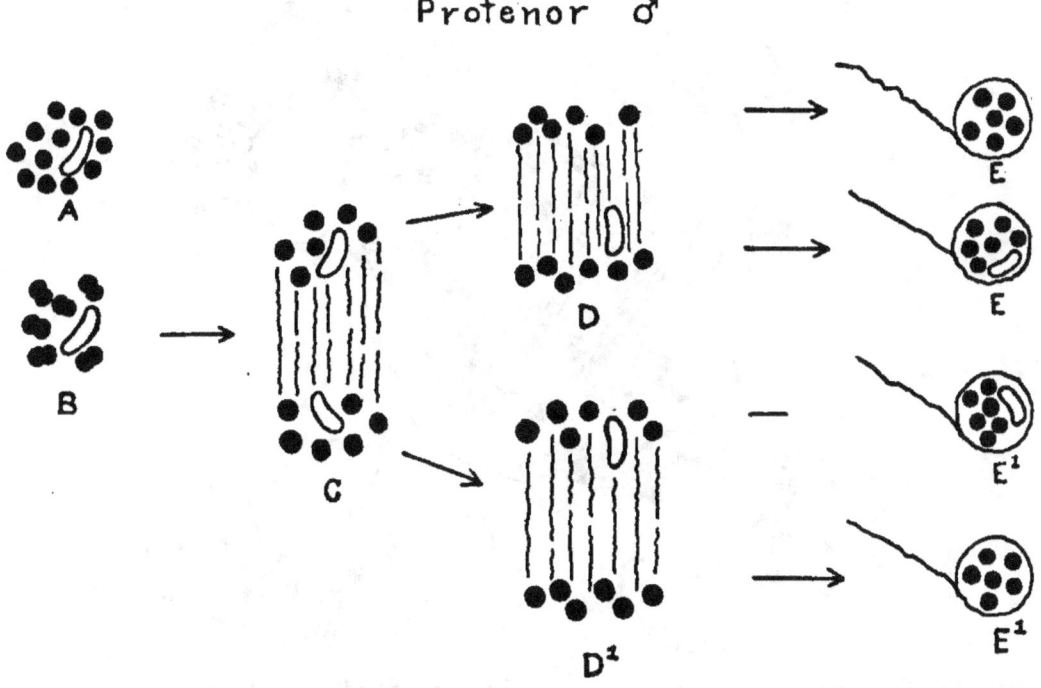

Fig. 14.—Diagram showing the distribution of the sex chromosome in *Protenor*. (After Morgan.)

among mammals. In general, animals in which this is the rule, bear both males and females, through all of the individuals must have been under the same environmental conditions. There are multiple births, however, in which the young are invariably of the same sex. Such is the case with those remarkably similar human twins known as *identical* twins. Such is the case with the four young in each litter of the nine-banded armadillo (Fig. 15). Now it can be shown that in these two and other similar instances, the several young are the product of a single fertilized egg which so develops as to form two or four complete individuals. If sex was determined after fertilization, one might expect a random sample of the two sexes here, but this is not the case.

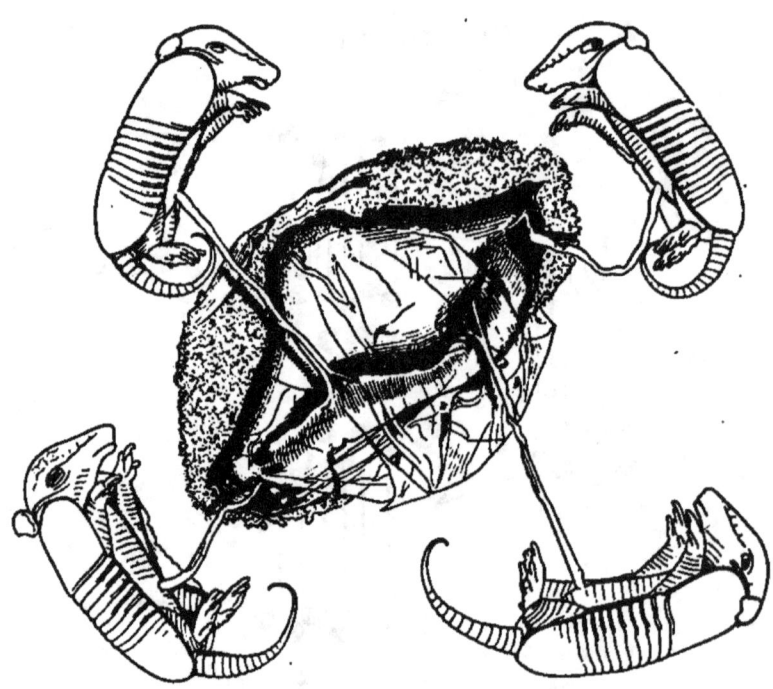

Fig. 15.—Identical quadruplets in the nine-banded armadillo. (After Newman from Doncaster.)

The chief support of this idea of sex determination, however, comes from the microscope and the breeding pen. In an ever increasing number of species, possibly including man himself, it has been found that besides the regular paired chromosomes, the autosomes, there is a single chromosome or possibly a chromosome group commonly known as the x-chromosome, whose behavior in cell division is somewhat different from the others, and whose distribution absolutely parallels the distribution of sex. There are two types. In the males of animals of Type A, which includes numerous flies, beetles, grasshoppers and bugs from among the insects, as well as representatives from several orders of mammals, a single x-chromosome is present in addition to the regular chromosome pairs, and for this reason *two* kinds of spermatozoa are produced at spermatogenesis in equal numbers, those possessing the extra element and those without it. In the females, on the other hand, *two* of these elements are present and the eggs, therefore, always possess it. Thus, on fertilization, half of the resulting young have *two* x-chromosomes and these become females, while half have but one and become males.

Diagrammatically it is this:

Ovum with x fertilized by sperm with x = female.
Ovum with x fertilized by sperm without x = male.

In some other cases (Type B), the eggs are dimorphic, while the sperm are all alike, but the result is the same; the sex distribution follows the chromosome differentiation.

In diœcious plants there is some evidence of a similar condition. Strasburger [200] found in one of the liverworts, Sphærocarpus, where the four spores produced by a single spore mother cell hang together and each such tetrad could be planted separately, that invariably two males and two females were produced. More recently Allen [1] has presented evidence of an x-chromosome in this genus. His discovery was made with material of the species *Sphærocarpus Donnellii*, but it has been corroborated by one of his students working with *Sphærocarpus texanus*.

Again, in the dioecious moss, Funaria, the Marchals,[130] by a remarkable series of regeneration experiments, have proven the determination of sex at the reduction division. Each spore was found to contain the potentialities of but one sex, but in the sporophyte they demonstrated the potentialities of both sexes by inducing direct aposporous development of gametophytes, which proved to have both antheridia and archegonia, the organs of both sexes.

The situation in hermaphroditic plants and animals is not so clear. Particularly in plants the peculiar life history with the introduction of alternation of generations, makes experimental work exceedingly difficult. Furthermore, there are many species of animals where the sex ratio is nowhere near equality and where both external and internal conditions undoubtedly do have marked influence, but in such a fundamental phenomenon we can hardly believe these difficulties are insurmountable or will lead to any radically different interpretation of the problem. Where there is such clear evidence from very different modes of attack and upon species so unrelated one is constrained to believe the obstacles to a unified theory are only superficial. This is particularly true since there is another line of experimental evidence in favor of the determination of sex by the chromosomes. Our whole evidence on inheritance, in fact, is linked up with chromosome distribution, so that the easiest way to visualize the process is by supposing that the individual potentialities, the *factors,* which coöperate in the development of plant and animal characters, are disposed in a definite manner in the chromosomes, as we shall see in the next chapter. The particular discoveries which demand our attention in connection with the phenomenon of sex, however, are those regarding characters commonly known as sex-linked, whose distribution can be accurately predicted if we assume they are definitely coupled with the sex determiner.

Such a character is hereditary color-blindness in man, a condition in which the affected individual cannot distinguish between red and green. It is far commoner in man than in women, and its inheritance is so peculiar that it often seems to skip a generation.

A color-blind man married to a normal woman will have only normal children of either sex. The sons will never have color-blind progeny by women with normal vision, but the daughters, though married to normal men, will transmit color-blindness to one-half of their sons. If, moreover, a daughter mates with a color-blind man, as might frequently happen in marriage between cousins, on the average one-half of her daughters as well as one-half of her sons will be abnormal.

This interesting and apparently complicated inheritance is really very simple if we merely assume that the sex chromosomes of the color-blind individuals also carry the determiner for color-blindness. Fig. 16 shows what would be expected. Representing the normal vision by boldface type and color-blindness by outline we see first the result of mating a normal woman with a color-blind man. Since all of her sex-cells, when matured, contain one *normal* x-element, and since the sex-cells of the male are of two kinds, half containing an abnormal or color-blind determining x-element and half containing no x-element whatever, it is obvious that the sons must receive their x-element only from their mother and the daughters must receive one of their x-elements from their father. The sons, therefore, cannot be color-blind and cannot transmit color-blindness, but the daughters, though

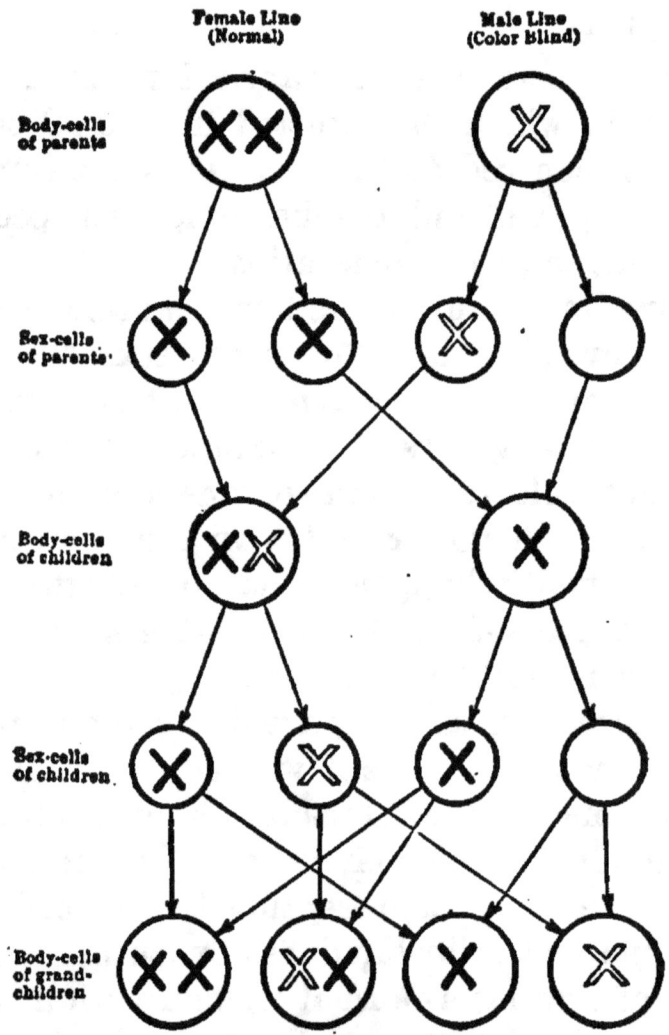

Fig. 16.—Diagram illustrating the inheritance of a sex-linked character such as color-blindness in man on the assumption that the factor in question is located in the sex chromosome. The normal sex chromosome is indicated by a black X, the one lacking the factor for color perception, by a light X. It is assumed that a normal female is mated with a color-blind male. (After Guyer. Courtesy Bobbs Merrill Co.)

they will not be color-blind themselves because one normal x-element is sufficient to determine normal vision, will produce *defective* x-elements in one-half of their ova,

THE MECHANISM OF REPRODUCTION

and for this reason will transmit color-blindness to one-half of their sons by a normal man, as will be seen by following out the fourth and fifth columns in the diagram. An egg containing the normal x-element can meet a spermatozoön carrying an x-element and thus produce a daughter, or it may meet a spermatozoön with no x-element and thus produce a son; but in either case the children will have normal vision. On the other hand, an egg containing a defective x-element will by similar fertilizations result either in a normal-visioned daughter, who will carry color-blindness in half of her ova, or in a son who will be *color-blind*.

Such a scheme of interpretation might seem quite visionary were it not for the fact that similar types of inheritance occur in many of the lower animals. By carefully controlled experiments with them it has been proven beyond a doubt.

CHAPTER IV

THE MECHANISM OF HEREDITY

The scientific era in the investigation of heredity began in the latter half of the nineteenth century with the work of Galton and of Mendel. Both enthusiastic and competent investigators, their efforts made with different material and from diverse points of view, did not fare the same. Galton measured the inheritance of groups of individuals by their resemblance to their progenitors and failed because his method could not take into account the true relationship between the germinal constitution and the body characters of an individual; Mendel determined the inheritance of a single organism by making the characters of its progeny the criterion and succeeded. Without knowledge of the cell mechanism of gametogenesis and fertilization, Mendel described the results of his hybridization experiments in terms which agreed precisely with these later discoveries in the field of cytology. Mendelian heredity has proved to be the heredity of sexual reproduction: the heredity of sexual reproduction is Mendelian.

Progress in the study of heredity through investigations patterned after Mendel's model has been so great that the subject now forms an important sub-division of general biology—Genetics. The details of the subject have outgrown the limits of a single volume, and a knowledge of the generalities is no longer confined to the professional biologist. For such reasons we propose to

THE MECHANISM OF HEREDITY

discuss here only the broader relationships of Mendelian heredity to the behavior of the chromosomes, since this phase must be emphasized as a basis for correlating the facts from Nature's experiments on inbreeding and outbreeding with the results from the experiments made by man.

The Mendelian method of studying heredity consists essentially in crossing forms which differ by well-defined characteristics and in following the distribution of these characteristics separately and quantitatively in the succeeding generations. If a wheat with a long lax head or spike is crossed with one having a short dense spike the F_1 (first filial) generation bears intermediate spikes. The F_1 generation, self-fertilized, however, yields all three types—long, intermediate and short spikes—in the F_2 generation; and in large numbers these types bear a constant ratio to each other in the proportion 1 long spike: 2 intermediate spikes: 1 short spike. Nor is this all. The long-spiked plants all breed true to long spikes, the short-spiked plants all breed true to short spikes, while the plants bearing intermediate spikes again produce the ratio exhibited by the F_2 generation. Diagrammatically the result of the cross is as follows:

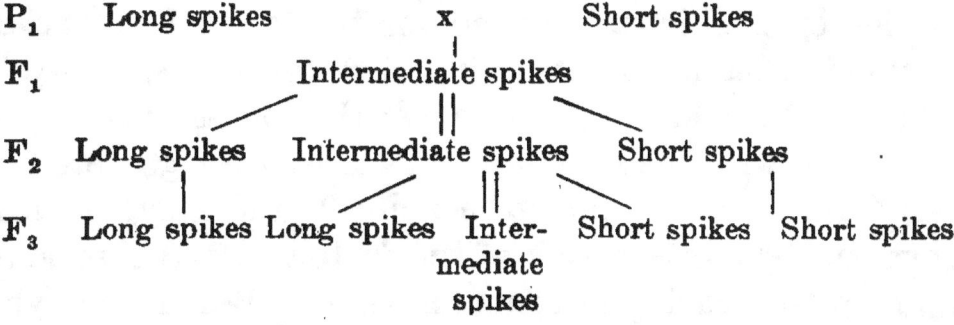

If the description of the dual nature of the cells of plants and animals and the result of gametogenesis is recalled, the reason for the production of the ratio of 1 long spike: 2 intermediate spikes: 1 short spike in the F_2 generation is plain. Furthermore, it is clear why the types like the grandparents breed true and the type like the hybrid F_1 generation does not breed true.

The long-spiked wheat has received the factor for long spikes, the something in the germ cell that stands for the production of long spikes, from *both* of its parents; there-

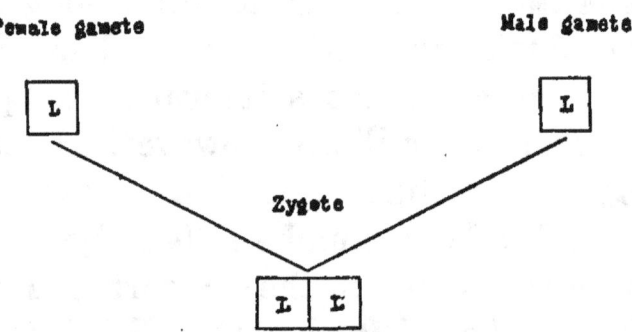

Fig. 17.—Diagram showing the union of like gametes.

fore it breeds true to long spikes. The gametes which it produces all bear the factor for long ears.

The diagram illustrating the fusion of the parental gametes, shows why the long-spiked wheat produces gametes, each of which bears the factor for long spikes. If the letter S is substituted for the letter L in the diagram, the same illustration holds for the short-spiked wheat. But what happens when the long-spiked variety is crossed with the short-spiked variety? A gamete bearing L fuses with a gamete bearing S and a zygote LS is formed. The interaction of the factors L and S produces an F_1 plant bearing intermediate ears. When this hybrid

THE MECHANISM OF HEREDITY 53

comes to produce gametes they bear either the one or the other—and never both—of these factors. In other words, the germ cells (both male and female) of the hybrid are half of them L and half of them S. When the F_1 generation is selfed, therefore, it is a matter of chance which of these germ cells meet to form zygotes. If a large progeny is produced, there will be a ratio of 1 $\boxed{L|L}$: 2 $\boxed{L|S}$: 1 $\boxed{S|S}$, and since the formulæ $\boxed{L|L}$ and $\boxed{S|S}$ are like the zygotic formulæ of the long-spiked and short-spiked parents, respectively, the plants that they produce will be long-spiked and short-spiked, as the case may be, and will breed true to that character. The intermediates, however, having been produced by zygotes $\boxed{L|S}$ like the F_1 generation, will behave in the same manner when selfed.

That the ratio will be approximately 1 $\boxed{L|L}$: 2 $\boxed{L|S}$: 1 $\boxed{S|S}$ is plain if one thinks for a moment what the result would be if a thousand tickets bearing the letter L and a thousand tickets bearing the letter S were shuffled up in a hat and drawn out in pairs, replacing the pair each time after drawing and recording. Suppose the first member of the pair represents the egg cell; the chances are ½ that it will be L or S. The second member of the pair represents the male cell and the chances are likewise ½ that it will be L or S. Therefore, when L is the first member of the pair, half of the time the zygote formed will be $\boxed{L|L}$ and half of the time it will be $\boxed{L|S}$. Likewise, when S is the first member of the pair, zygotes $\boxed{S|L}$ and $\boxed{S|S}$ will be formed in equal quantities. Combining these possibilities, the ratio 1 $\boxed{L|L}$: 2 $\boxed{L|S}$: 1 $\boxed{S|S}$ is obtained. Diagrammatically the expected production of gametes and zygotes is as follows:

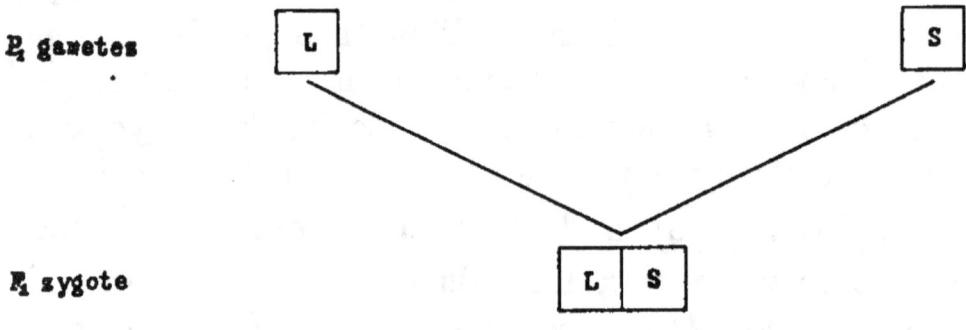

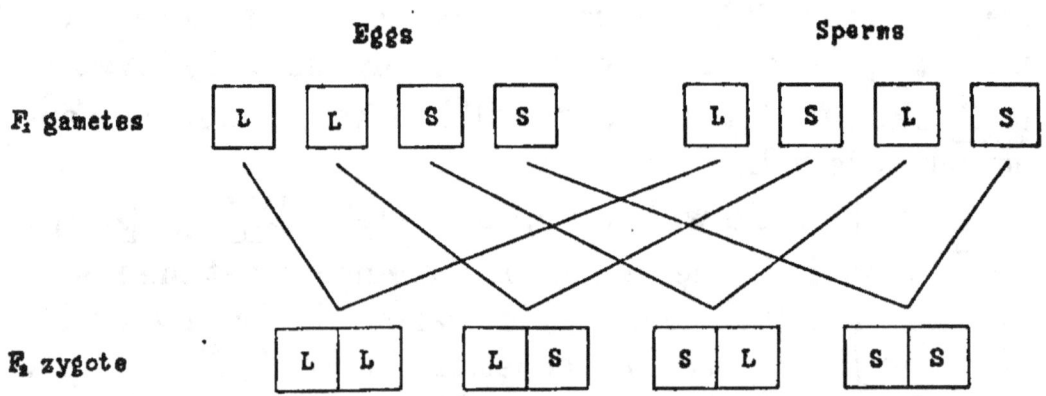

Fig. 18.—Diagram to illustrate *Mendelism* in a cross between long-spiked and short-spiked wheat.

Let us express the whole matter mathematically. The chance of an event happening in an infinite number of trials is expressed by a fraction of which the numerator is the number of favorable ways and the denominator the whole range of possibilities, both favorable and unfavorable, if each is equally likely to occur. Hence, certainty is expressed in the figure 1. Furthermore, the chance that two or more independent events will happen together is the product of their respective chances of happening. The chance of an L egg meeting an L sperm is $½ × ½ = ¼$, and the chance of an L egg meeting an S sperm is $½ × ½ = ¼$. Similarly, the chance of an S egg meeting an L sperm is $¼$ and of an L egg meeting an S sperm $¼$. The sum of the possibilities is then

$$L + L = 1/4$$
$$S + L = 1/2$$
$$S + S = 1/4$$

One important thing to be remembered, however, is that the law of chance expresses the probability when infinitely large numbers are concerned. When small numbers are dealt with, departures from the expected ratios are obtained. It is a matter of common sense that although the chances of throwing heads with a penny are $½$, yet in a small number of throws exactly $½$ of them may not be heads. A regular ratio of these departures is to be expected which accords with the law of error.

Typically, experiments such as this are the basis for Mendel's Laws of Inheritance, the first of which may be stated as follows: *Unit factors contributed by two parents having definite rôles in the development of characters, separate in the germ cells of the offspring without having influenced each other.* Although Mendel himself knew of no mechanism by which such a process could take place, although his theory was evolved wholly as a fitting interpretation of the facts obtained by breeding, what can be more reasonable than to suppose that the germ cell factors reside in the chromosomes and that the separation of the chromosomes at the reduction division in gametogenesis furnishes the segregation of factors required?

The interaction of a pair of homologous factors in a hybrid—allelomorphs they are called—does not always result in the production of an intermediate. Often the action of one factor *dominates* the action of the other, either by masking it or by inhibiting its operation. When this occurs the dominated character recedes from sight in the F_1 generation and the ratio in the F_2 generation is 3 dominant : 1 recessive. But since only one-third of these dominants breed true [a] and two-thirds behave as did the F_1 generation, the results are, therefore, comparable with those illustrated by the wheat, and the phenomenon of dominance is a mere detail.

Mendel was not content with experiments in which only one pair of differentiating characters was concerned. He made crosses between varieties of the garden pea which differed by two and by three allelomorphic pairs of characters, and was rewarded for his perseverance by discovering a second law, usually known as the Law of Recombination. *This law states that two or more allelomorphic pairs of factors may segregate independently and may recombine in all the combinations possible governed by chance only.*

Though thousands of characters have been investigated since Mendel's time one cannot improve on the original classic as illustrative material for explanation of dihybrid heredity.

With two pairs of characters he designated the factors representing the dominant characters as A and B and the factors representing the recessive characters as a and b. The characters in the varieties crossed were as follows:

Seed parent $\begin{cases} A \text{ form round} \\ B \text{ cotyledon yellow} \end{cases}$ Pollen Parent $\begin{cases} a \text{ form wrinkled} \\ b \text{ cotyledon green} \end{cases}$
AB ab

---

[a] When the factor for a character has been received from both parents the organism is said to be *homozygous* for it; if it has been received from only one parent the individual is *heterozygous* or hybrid for it.

When these two forms were crossed all of the hybrid's seeds appeared round and yellow (AB) like those of the seed parent; that is, the round character and the yellow character were each dominant.

When these F_1 seeds were sown and the plants self-fertilized, four kinds of seeds appeared in the progeny in the four combinations that were possible, with the following numbers of each:

AB	round and yellow	315
aB	wrinkled and yellow	101
Ab	round and green	108
ab	wrinkled and green	32

These figures stand approximately in the relation of $9AB$ to $3aB$ to $3Ab$ to $1ab$. The forms appeared to belong to but four homogeneous classes (phenotypes). This was due to the phenomenon of dominance masking the difference between homozygotes and heterozygotes. By their behavior in the next generation they were found to belong to nine really different classes.

From the round yellow seeds (apparently AB) were obtained by self-fertilization:

(1)	$AABB$	seeds all round and yellow	38
(2)	$AABb$	seeds all round, yellow and green	65
(3)	$AaBB$	seeds all yellow, round and wrinkled	60
(4)	$AaBb$	seeds round and wrinkled, yellow and green	138

From the round and green seeds (apparently Ab) were obtained:

(5)	$AAbb$	seeds all round and green	35
(6)	$Aabb$	seeds all green, round and wrinkled	67

From the wrinkled and yellow seeds (apparently aB) were obtained:

(7)	$aaBB$	seeds all wrinkled and yellow	28
(8)	$aaBb$	seeds all wrinkled, yellow and green	67

From the wrinkled and green seeds (apparently ab) were obtained:

(9)	$aabb$	seeds all wrinkled and green	30

The total number of plants in the F_3 generation is not quite the same as the F_2 generation due to seed not germinating or plants dying, but it is plain that the ratio of $9AB : 3aB : 3Ab : 1ab$ obtained in the F_2 generation is made up of the following actual classes:

9 apparently AB made up of
$\begin{cases} 1 \ AABB \\ 2 \ AaBB \\ 2 \ AABb \\ 4 \ AaBb \end{cases}$

3 apparently Ab made up of
$\begin{cases} 1 \ AAbb \\ 2 \ Aabb \end{cases}$

3 apparently aB made up of
$\begin{cases} 1 \ aaBB \\ 2 \ aaBb \end{cases}$

1 apparently ab made up of $\quad 1 \ aabb$

The four classes $AABB$, $aaBB$, $AAbb$ and $aabb$, having each factor in the duplex condition bred true; the heterozygous condition of one or both allelomorphs in the remainder was shown by the character of the F_4 progeny.

In order to visualize these facts, or in truth the facts from any number of pairs of allelomorphs in independent Mendelian inheritance, one has only to recall that pairs of homologous chromosomes—one paternal and one maternal—meet during the process of gametogenesis and *one or the other* of each pair passes to either daughter cell. The factors A and a lie in corresponding loci in one pair of chromosomes, the factors B and b lie in corresponding loci in a second pair from among the seven paired chromosomes of the garden pea. Thus gametes bearing the factors AB, Ab, aB and ab will be formed in equal quantities in both the egg cells and the pollen cells, as is shown by the accompanying diagram.

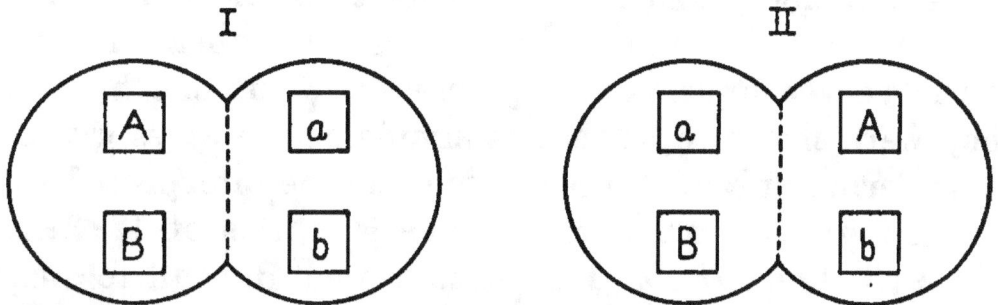

FIG. 19.—Diagram to illustrate gamete formation in a dihybrid in independent inheritance.

It is clear that if this process occurs in both male and female germ cells, and these germ cells unite by chance, the result is that obtained in the breeding experiments. The posssible matings can be shown graphically as follows:

	SPERMS			
Eggs	A B	A b	a B	a b
A B	A B A B	A B A b	A B a B	A B a b
A b	A b A B	A b A b	A b a B	A b a b
a B	a B A B	a B A b	a B a B	a B a b
a b	a b A B	a b A b	a b a B	a b a b

A large series of results on independent Mendelian heredity obtained during the past eighteen years by numerous biologists can be interpreted in the same manner. At first glance many of their results appeared to be either very complex or very irregular, but one by one they were shown to be just as simple as the cases given. Considering only instances which may be interpreted by two factors, for example, we have F_2 ratios of 12:3:1, 9:7, 9:3:4 and 13:3. But these are not difficult to analyze. They are simply the ordinary 9:3:3:1 ratios in which part of the terms are combined in various ways. For example, the F_2 ratio, when certain black beans are crossed with certain white beans, is 12 black : 3 yellow : 1 white. Clearly this is because black + yellow (AB) is in appearance not different from black (A) alone. A purple sweet pea crossed with a certain white variety segregates in a ratio of 9 purple : 7 white. This result is easily explained by assuming that the purple color only appears when a color base factor, A, is present in connection with a color-producing factor, B. The last three terms of the dihybrid ratio, $3AB : 3aB : 1ab$, therefore, are alike in appearance. This assumption has been proved to be correct in other ways. Again if a black variety of rat, mouse, guinea-pig or rabbit be crossed with a white variety carrying a factor for ticking the hair with yellow, known as agouti, the segregating ratio is 9 agouti : 3 black : 4 white. The reason for the combination of classes Ab and ab is because the agouti factor does not show except in the presence of color (B). Finally, a factor A which inhibits the action of B and therefore makes AB and Ab resemble ab, gives the peculiar ration 13:3. Crossing a certain race of white fowls with colored races, for instance, gives the ratio 13 white : 3 colored.

It is evident, if one runs over these examples and works out all the possibilities involved, he will find that two white races of sweet pea, when crossed, will give purples in the F_1 generation, a white race of guinea-pigs, crossed with a black variety, will give all agouti, etc. Such curious results are actually obtained. They are quite simple, and their whole heredity may be visualized by the use of the same chromosome scheme as given above. Of course, some of them require the assumption of differences in more than two allelomorphic factors, but this can be done by remembering that additional factor pairs follow the same mathematical scheme as do one or two pairs.

No matter how satisfactory it would be to have all biological facts interpreted with a primer simplicity, the truth is that animals and plants are complex organizations. Probably only the tiniest fraction of the germ cell constitution of any organism has ever been analyzed through Mendelian methods, yet in the pomace fly *Drosophila melanogaster*, in which there are only four pairs of chromosomes, Morgan and his associates have traced the hereditary transmission of well over one hundred factors, each of which has one or more functions to perform in the development of characters in the adult. It is obvious that with such a large number of characters and such a small number of chromosomes, a single chromosome must carry many factors. This conclusion granted, it would seem as if any of these groups of factors carried by a single chromosome would necessarily behave as single factors; in other words, they would enter a cross in a group and be segregated in a group in the F_2 generation.

Such cases have appeared in breeding experiments, but they are very rare and are probably not what they seem, because of an insufficient number of individuals from which to draw conclusions. What usually happens is for these sets of factors to *tend* to hang together at the reduction division in the F_1 generation. They tend to be *linked,* but the linkage is often broken.

An example from Morgan's work on the pomace fly will make this clear. If a female fly with black body and vestigial wings be crossed with a wild male having a gray body and long wings, the result is offspring like the male, gray body and long wings being dominant. Now, since these F_1 individuals have one chromosome containing the factors for black body and for vestigial wings, and a homologous chromosome containing the factors for gray body and for long wings, one would expect gametes of only two kinds to be formed at the maturation of the eggs and sperms. If this were true, an F_2 generation obtained by mating a male and a female from the F_1 generation should consist of 3 flies having long wings and gray bodies to 1 fly having vestigial wings and a black body. But this is not the result obtained. In addition to large numbers of flies of this type, there are smaller numbers of flies characterized by long wings and black bodies, and by vestigial wings and gray bodies. Such a result, on a chromosome basis, could only be obtained through the homologous chromosomes interchanging their factors at the reduction division.

There is good cytological evidence that such an interchange of chromosome parts does take place at gametogenesis. At the time when the homologous pairs of chromosomes approach each other just previous to the reduction of the chromosome number to half, they twist around each other. Often they retain their individuality as they pass to the daughter cells; but sometimes they break at various places, join their parts in a different combination and pass to the daughter cells in their new guise. The diagram will make this clear.

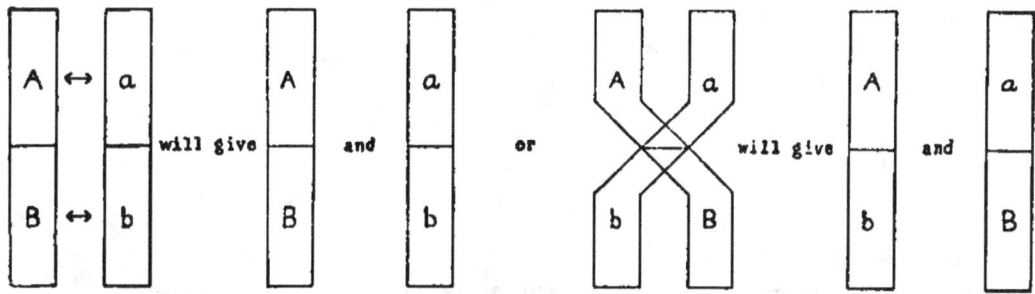

Fig. 20.—Diagram to illustrate gamete formation in a dihybrid in linked inheritance.

The easiest way to determine the frequency with which these breaks in linked characters occur, and a way which gives it in terms of chromosome *crossovers* is to mate F_1 individuals back to the double recessive type. When the F_1 male in the cross just described is mated with a black vestigial female, only two classes of offspring are produced; half are black vestigial and half are gray long in type. The F_1 male produces only two kinds of gametes. There is, therefore, no crossing over between the chromosomes of the male.

On the other hand, when the F_1 female is mated with a black vestigial male, four types of offspring are produced:

Non-crossovers		Crossovers	
Black vestigial	Gray long	Black long	Gray vestigial
41.5 per cent.	41.5 per cent.	8.5 per cent.	8.5 per cent.

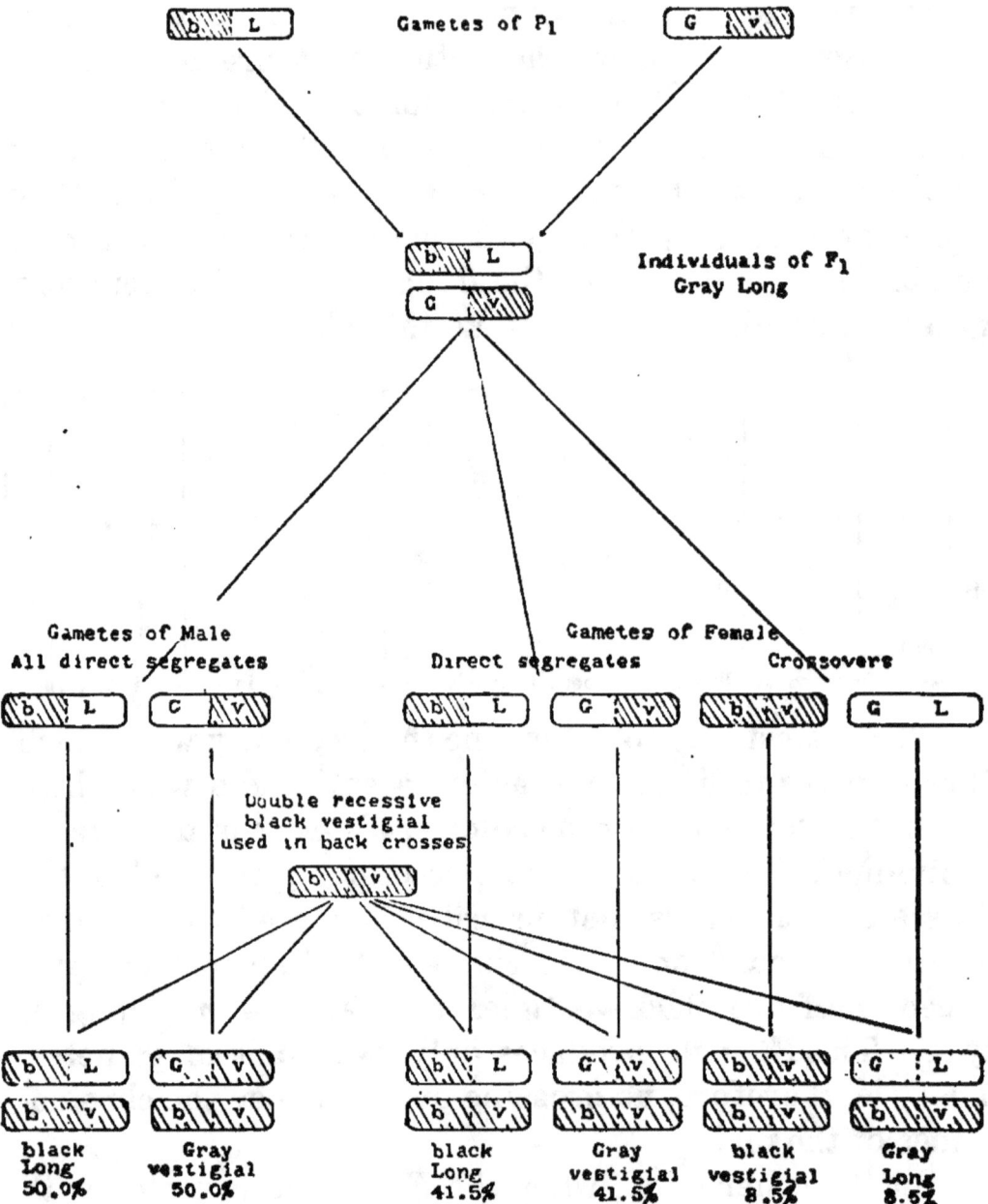

Fig. 21.—Diagram to illustrate linkage breaks or "crossing-over" between the loci containing the factors for the allelomorphic pairs. Gray body, black body and long wing—vestigial wing in Drosophila. (After Morgan.)

A complete visualization of this series of matings is given in Fig. 21.

The peculiar fact that there is no crossing over in the males need not concern us here. In other species there is evidence of crossing over in both sexes. What is important is that crossing over does occur with a definite frequency and this frequency is constant for any particular pair of characters except when modified in various *ways* which can be given concrete explanations. It does not matter, moreover, whether the two factors enter the cross as $\genfrac{}{}{0pt}{}{A|a}{B|b}$ or as $\genfrac{}{}{0pt}{}{A|a}{b|B}$, crossing over is the same in each case; that is, the tendency is just as great for Ab to stay together after they are in that combination as for AB to stay together when that particular combination characterizes the individual.

The gross result of several thousand experiments of this character, therefore, is to associate Mendelian inheritance more definitely than ever with the chromosomes. What seemed to be an exception, furnished striking evidence in support of the theory. It is true, a few isolated instances of characters which appear to be carried by cell substances other than the chromosomes have been discovered; but it is pretty clear that in the main all of the varied characters which differentiate the individuals within a single species—and these are the only ones which can be studied profitably through crosses—are controlled by the distribution of factor units which lie within the chromosomes.

We can visualize the whole process of heredity by means of chromosome diagrams just as we can visualize the whole process of chemical recombination through models of the atoms constructed to fit the facts furnished by chemical reactions. The result is that we can do much toward predicting what will happen under given conditions because we know what has happened under similar conditions. For example, it having been determined that in the pomace fly many characters are linked to chromosomes whose distribution parallels that of sex, we know it to be much more than a guess to say that the color-blindness of man of which the hereditary distribution was described in Chapter III, is controlled by a factor lying in the sex chromosome and recessive to the normal.

Though the whole mechanism in the higher plants and animals can thus be pictured as one of sexual reproduction, in its details the results are still too complex to analyze as concretely as the cases given for illustration. Several thousand concrete differences between plants of the various angiosperm families and between animals in at least three different phyla have been followed through pedigree cultures sufficiently carefully to make possible a definite factorial analysis of their hereditary transmission. This has been possible, first, because variation has taken place in these factors, enabling one to follow the transmission of each member of an allelomorphic pair, and, second, because this variation has been somewhat qualitative in nature. Unfortunately for the peace of mind of the biologist, however, the more numerous differences between animals and between plants are the quantitative differences, the variations which make organs a little larger or a little smaller. Now it is a great deal easier to determine the transmission of the factor differences which determine that one flower shall be red and another white than it is to trace the distribution of the factors which determine that one flower shall be one inch and another two inches long. Nevertheless, through the efforts of numerous investigators it has been possible to show that such hereditary differences behave as should be expected if their inheritance follows the same laws as do the simpler characters. The basis, as one might say, of the Mendelian interpretation of size differences is the proof that practically all qualitative characters are affected by numerous factors. Sometimes there are two or more factors which produce nearly identical visible results, but more often the character complex is affected in different ways and in various degrees by particular factors. Whether the character develops at all or not seems to be due to the presence or absence of one or more main factors, but given the presence of these factors the degree of development may be influenced by many subsidiary factors or modifiers. Now these modifiers being transmitted independently of one another and of the principal factor or factors, an individual carrying certain modifiers and lacking the principal factor may be crossed with an individual carrying the main factor and lacking the modifiers. The result is a series of recombinations among the germ cells of the F_1 generation which produces F_2 individuals carrying various groups of modifiers and therefore developing the character complex under consideration in different degrees.

If one studies carefully such crosses as the one just described, he finds that a number of general conditions are fulfilled.

1. When pure or homozygous races are crossed, the F_1 populations are similar to the parental races in uniformity. This conclusion devolves from observations that if any particular factors AA and aa are homozygous in the parental races, they can only form Aa individuals in the F_1 generation.

2. If the parental races are pure, F_2 populations are similar, no matter what F_1 individuals produce them, since all variability in the F_1 generation is the result of varying external conditions.

3. The variability of the F_2 populations produced from such crosses should be much greater than that of the F_1 populations, and if a sufficient number of individuals are produced the grandparental types should be recovered. The fulfillment of this condition comes about from the general laws of segregation of factors in F_1 and their recombination in F_2.

4. In certain cases F_2 individuals should be produced showing a greater or a less extreme development of the character complex than either grandparent. This is merely the result of recombination of modifiers, as was explained above.

5. Individuals of different types from the F_2 generation should produce populations differing in type. The idea on which this statement is based is, of course, that all F_2 individuals are not alike in their inherited constitution and therefore must breed differently.

6. Individuals either of the same or of different types chosen from the F_2 generation should give F_3 populations differing in the amount of their variability. This conclusion depends on the fact that some individuals in the F_2 generation will be heterozygous for many factors and some heterozygous for only a few factors.

Such are the conditions which must be fulfilled by crosses exhibiting size differences if we are to visualize their inheritance in the same way as we visualize the inheritance of qualitative characters such as color. If the size differences are controlled by numerous germ-cell factors, the distribution of the latter cannot be followed with the same ease as one would follow the distribution of cotyledon colors in the garden pea. This is true because the visible effects of certain factors is sure to be very small, and because varying external conditions obscure the effects of inheritance. For example, a plant which through its inheritance should become 6 feet tall under average conditions may become only 4 feet tall if planted in a sterile soil, but a plant which under average conditions would become only 4 feet tall might become 5 feet tall if grown in a very fertile soil.

Nevertheless, in spite of these drawbacks, one can select size characters for study which are influenced but slightly by external conditions, and by studying large numbers through several generations, and by applying mathematical tests to determine the uniformity or the variability of the resulting populations, he can find out whether quantitative characters satisfy the six requirements seen to be fulfilled by qualitative characters. This has been done in numerous cases, and the results firmly convince all unprejudiced investigators that the inheritance of all types of characters is the same.

Table I, from crosses between two varieties of *Nicotiana longiflora*[54] differing in the size of their flowers, illustrates the point. One does not need any refined mathematical methods to see that when the small variety having flowers about 40 mm. in length is crossed with the large variety having flowers about 94 mm. in length; the result is a uniform F_1 population having flowers about 64 mm. in length. The two F_2 populations which it produced are much more variable; and one can easily calculate that if several thousand plants had been grown instead of

TABLE I

Frequency Distributions for Corolla Length in a Cross Between Varieties of Nicotiana Longiflora Cav.

Designation	Year	Generation	Parent size	34	37	40	43	46	49	52	55	58	61	64	67	70	73	76	79	82	85	88	91	94	97	100
No. 383	1911	—	—		13	80	32																			
No. 383	1912	—	—	1	4	28	16																			
No. 383	1913	—	—		4	32	1																			
No. 330	1911	—	—																			6	22	49	11	1
No. 330	1912	—	—																			2	16	32	6	
No. 330	1913	—	—																			5	7	10	2	
No. 383×330	1911	F$_1$	61							4	10	41	75	40	3											
No. (383×330) 1	1912	F$_2$	61					1	5	16	23	18	62	37	25	16	4	2	2	1						
No. (383×330) 2	1912	F$_2$	61					2	4	24	37	31	38	35	27	21	5	6	1							
No. (383×330) 1-1	1913	F$_3$	72											4	20	25	59	41	19	2						
No. (383×330) 1-2	1913	F$_3$	46			1	4	26	44	38	22	7	1													
No. (383×330) 1-3	1913	F$_3$	50			1	6	20	53	49	15	4														
No. (383×330) 1-4	1913	F$_3$	60				2	3	9	25	37	70	19	10												
No. (383×330) 2-1	1913	F$_3$	77						1	0	1	1	2	16	33	43	34	20	6	1						
No. (383×330) 2-2	1913	F$_3$	81											1	1	8	16	20	32	41	17	3	3	1		
No. (383×330) 2-3	1913	F$_3$	80											2	8	14	21	39	32	10	1					
No. (383×330) 2-4	1913	F$_3$	50						7	25	55	55	18													
No. (383×330) 2-5	1913	F$_3$	82												3	5	12	20	40	41	30	9	2			
No. (383×330) 2-6	1913	F$_3$	44			8	42	95	38	1	1															
No. (383×330) 1-2-1	1914	F$_4$	43			2	23	122	41	1																
No. (383×330) 1-3-1	1914	F$_4$	85															4	9	38	75	59	6	3	1	
No. (383×330) 2-6-1	1914	F$_4$	87												4	5	6	11	21	33	41	29	8	5	1	
No. (383×330) 2-6-2	1914	F$_4$	41	3	6	48	90	14																		
No. (333×330) 1-3-1-1	1915	F$_5$	41																							
No. (388×330) 2-6-2-1	1915	F$_5$	90													2	3	8	14	20	25	25	20	8		

NOTE. The numbers shown enable one to trace relationships. For example, 1-1 and 1-2 are F$_3$ populations from 1 in F$_2$, 2-1 and 2-3 are F$_3$ populations from 2 in F$_2$.

THE MECHANISM OF HEREDITY

about 200, the grandparental sizes probably would have been obtained. Furthermore, if one studies the results obtained in the F_3, F_4 and F_5 generations, considering only the range of their variability, it is clear they differ in both type and extent of variation.

No assumptions unproved for the inheritance of qualitative characters are necessary for thus visualizing the inheritance of quantitative characters, and no facts discovered in tracing the inheritance of other characters—such as those involving linkage—are overlooked. But in order to picture the situation easily, let us assume that dominance is usually absent (often the case), that two doses (*i.e.*, the homozygous condition) of a factor have twice the effect of one dose (true for all practical purposes), that independent factors cumulative in their operation are allelomorphic to their absence in the hybrid (linkage though it complicates matters, does not change our reasoning).

Let us assume a case of "blended" inheritance where all fluctuations due to environment are eliminated. A plant 12 inches tall is supposed to be crossed with a plant 28 inches tall. The difference between them is 16 inches. If this difference is due to one allelomorphic pair in which dominance is absent, the F_1 generation is all intermediate —about 20 inches—and the F_2 generation falls into three classes in which two represent the grandparental forms and one represents the F_1 form. Twenty-five per cent. are 12 inches tall, fifty per cent. are 20 inches tall and twenty-five per cent. are 28 inches tall.

But suppose this 16-inch difference between the parents is represented by two allelomorphic pairs instead of one. The F_1 generation is again 20 inches tall, but instead of there being three classes in F_2, there are five classes, *viz.*, 12, 16, 20, 24 and 28 inches, and they appear in the ratio 1:4:6:4:1. Each grandparental type appears once out of sixteen times.

The way this ratio is obtained is by simple recombination, but as dominance is absent, each time a *single* "presence" factor is added, the height is increased four inches.

9	1 *AABB*	=	28 inches
	2 *AaBB*	=	24 inches
	2 *AABb*	=	24 inches
	4 *AaBb*	=	20 inches
3	1 *AAbb*	=	20 inches
	2 *Aabb*	=	16 inches
3	1 *aaBB*	=	20 inches
	2 *aaBb*	=	16 inches
1	1 *aabb*	=	12 inches

If three independent size characters instead of two were involved in this cross, the F_1 individuals would fall in the same class as before, but the F_2 classes would be seven in number and the grandparental sizes would each be recovered only once out of sixty-four times. For four factors there would be nine classes of F_2 individuals, and the grandparental types would each occur only once out of two hundred and fifty-six times; while with only eight factors, the forms of the grandparents would each appear only once out of 65,536 times, and it would be quite remarkable if they were ever recovered from an ordinary cross.

The entire scheme of this type of inheritance can be expressed in mathematical form just like ordinary Mendelian inheritance with full dominance. Let us recall that

THE MECHANISM OF HEREDITY

the F_2 Mendelian expression for N allelomorphic pairs when dominance is complete is the expanded bionominal:

$(3+1)^n$ or $(3/4+1/4)^n$
$N=1 \quad (3+1)^1 = 3+1$
$N=2 \quad (3+1)^2 = 3^2 + 3 + 3 + 1 = 9 + 3 + 3 + 1$
$N=3 \quad (3+1)^3 = 3^3 + 3(3^2)\ 2 + 3\ (3) + 1 = 27 + 9 + 9 + 9 + 3 + 3 + 3 + 1$

Likewise, the expanded bionomial $(½ + ½)^{2n}$ gives the numerical relationships when dominance is absent and N represents the number of allelomorphic pairs. The expression is $(½ + ½)^{2n}$ instead of $(½ + ½)^n$ because it is supposed that the presence of any allelomorphic pair in the heterozygous condition produces one-half the visible effect on the character that is produced when the hereditary factors are present in the homozygous condition. When N is very large the frequencies with which the different classes occur form a regular curve called the normal curve of error. This is the curve that is produced when the errors in any physical measurement are similarly plotted, using as classes any constant deviation from the average, as a, $2a$, $3a$, etc. This same curve is also produced when one plots the fluctuations of many organic characters produced by the infinite complexity of external conditions.

If no non-heritable fluctuations intervened to obscure the class to which any particular zygote belongs, therefore, one should expect the following classes in F_2 when parents of different sizes differing in N allelomorphic pairs are crossed. The extremes represent the grand-parental types in each case, and the intermediate classes theoretically divide the difference between the parents into aliquot parts. It should be noted, however, that this is theory only; in reality the influence of one factor may be somewhat different from that of another factor.

$N=1$					1	2	1			
$N=2$				1	4	6	4	1		
$N=3$			1	6	15	20	15	6	1	
$N=4$		1	8	28	56	70	56	28	8	1
$N=5$	1	10	45	120	210	252	210	120	45	10	1

Let us now note a few of the practical difficulties in interpreting results that may follow this method of inheritance. In the theoretical example that we have used for the sake of clearness, it was assumed that there were no non-heritable fluctuations due to environment. Unfortunately this is not the case in nature. Fluctuations are everywhere present. They would obscure the classes to which individuals belong even if these class differences were quite large. And since they are usually small, the change of individual form due to environmental causes makes it impossible to separate an F_2 population into the true classes to which they belong gametically. Nor is this the whole trouble. If the table showing the expected results with two pairs of size characters is examined, it is found that not all the individuals that belong to a particular size class have the same zygotic formula. For this reason one cannot pick out zygotes of a certain size and expect them to breed the same. Their potentialities are likely to be different. Furthermore, practical breeding results are undoubtedly complicated by cases of correlation. This correlation need not be gametic, though such cases in all likelihood do occur; it may be merely physiological. For example, a maize plant might have the gametic possibilities of small plant size and large ear size, but it would be foolish to expect that a plant capable of only a limited amount of development could bear as large an ear as if it were as a whole capable of greater size development. Thus it must not be expected that theoretical possibilities are always expressed perfectly in nature, any more than it should be expected that theoretical physical calculations concerning known laws should agree perfectly with experimental data. Plants and animals do indeed seem to have in their reproductive cells a mosaic of independently transmissible factors, but a plant or animal is certainly not to be described as a mosaic of independent unit characters. These factors that appear to be independent in heredity act and react upon one another in complex ways during their development.

Hundreds of studies on quantitative characters have been made. They all have the same result. Mendelian inheritance rules. Plants appear to be less complex than animals. A size complex in an animal seems to be the result of the interaction of a large number of factors, a size complex in a plant appears to be the result of the interaction of a small number of factors. But the mode of inheritance is always the same. It is the result of the behavior of the chromosomes, and one can picture it with the greatest ease with the simple diagram of the reduction division at gametogenesis if he fancies to himself that the chromosomes are carrying bodies for the unit factors of heredity, that the arrangements within them are a near approach to perfection, that exchanges of contents may be made only with regularity and precision so no essential feature of the mechanism shall break down.

In thus visualizing the process of heredity, one must not be so overcome by the beauty of the picture that he is unable to realize just what has been done. He must not forget which part of this diagrammatic representation of the heredity mechanism is fact and which part is theory, for confusion between the two has led to a regrettable controversy over a point which is of paramount importance in any discussion of inbreeding and outbreeding —the stability of inherited factors.

The relation between fact and theory in the Mendelian conception of inheritance is this: Various kinds of animals and of plants were crossed and the results recorded. With the repetition of experiments under comparatively constant environments these results recurred with sufficient regularity to justify the use of a notation in which theoretical factors or genes located in the germ cells replaced the actual somatic characters found by experiment. Later, the observed behavior of the chromosomes justified localizing these factors as more or less definite physical entities residing in them. Now the data from the breeding pen or the pedigree culture plot and the observations on the behavior of the chromosomes during gametogenesis and fertilization are facts. The factors are part of a conceptual notation invented for simplifying the description of the breeding facts in order to utilize them for purposes of prediction, just as the chemical atom is a conception invented for the purpose of simplifying and making useful observed chemical phenomena. As used mathematically, both the genetical factor and the chemical atom are concepts, but biological data lead us to believe that the term factor represents a biological reality of whose nature we are ignorant, just as a molecular formula represents a physical reality of a nature yet but partly known.

With this distinction in mind, one may treat the factor—or the atom—from two points of view, either as a mathematical concept or a physical reality. As a mathematical concept it is the unit of heredity, and a unit in any notation must be stable. If one creates a hypothetical unit by which to describe phenomena and this unit varies, there is really no basis for description. He is forced to hypothecate a second fixed unit to aid in describing the first.

The point at issue in this connection may be explained as follows: Characters do vary from generation to generation, and the question to be decided is, how much of this variation is due to the recombination of factors (considered now as physical entities) and how much is due to change in the constitution of the factors themselves. The obvious way to determine such a matter is first to appeal to Nature and see whether it is possible for characters to have a long period of stability under any conditions; and, second, to investigate the stability of characters when the environment is comparatively constant and change due simply to recombinations of heterozygous factors is eliminated.

Of the results of the appeal to Nature only one need be mentioned. Wheeler [215] has found that ants preserved in amber of the Oligocene period, fossils which are better preserved than any others, and which are thought to be at least 3,000,000 years old, are practically identical with living species. The only points of variance to be observed are slight differences in shade of color, something probably due to the mode of preservation. Thus it is clear that organic characters may remain stable for periods of time so great as to be beyond our powers of realization.

Investigations as to the effect of selection on homozygous hermaphroditic plants which are self-fertilized naturally—the only material having a critical value with this mode of attack, if we except unicellular organisms reproducing asexually—have been made by a number of biologists following the lead of Johannsen,[109] who opened up the possibilities of this type of experiment. The results have always been the same. Characters are remarkably stable. They do change, but they change so rarely that a more useful purpose is served by identifying the physical unit factor with the mathematical factor unit, than to assume without justification that the physical factor is constantly changing and must be described by complex mathematical formulæ using other hypothetical units having no warrant for a physical existence. It is true, two investigations by Jennings[107] and by Middleton[142] have shown a seemingly more unstable condition in the infusorians *Difflugia coronata* and *Stylonychia pustulata*. But there are several reasons for not believing conclusions derived from data on these animals applicable to the higher plants and animals in which our real interest lies, without mentioning several technical points which might lead to interpretations different from those given by the authors. First, they are cases of asexual, not sexual, reproduction. Second, the germ plasm of the infusoria may not be insulated from the effects of environment as is the germ plasm of the higher organisms. Third, measurable differentiation in these experiments sometimes took such a number of generations that in man it would take some 3000 years to produce like results.

For these and other reasons which might be given, could further space be devoted to the subject, we believe there should be no hesitation in identifying the hypothetical factor unit with the physical unit factor of the germ cells. Occasional changes in the constitution of these factors, changes which may have great or small effects on the characters of the organism, do occur; but their frequency is not such as to make necessary any change in our theory of the factor as a permanent entity. In this conception biology is on a par with chemistry, for the practical usefulness of the conception of stability in the atom is not affected by the knowledge that the atoms of at least one element, radium, are breaking down rapidly enough to make measurement of the process possible.

CHAPTER V

MATHEMATICAL CONSIDERATIONS OF INBREEDING

The term *inbreeding* can be used in a relative sense only, except when dealing with hermaphroditic organisms. To say that one individual of a bisexual species is inbred and another not is as indefinite as saying one is short, the other tall. Strictly speaking, inbreeding refers only to the way in which individuals are mated together. This fact is well expressed by Pearl,[173] who says: " It is clear that underlying all definitions of inbreeding is to be found the concept of a *narrowing* of the network of descent as a result of mating together at some point in the network of individuals genetically *related* to one another in some degree. Let us take this as our basic concept of inbreeding. It means that the number of potentially *different* germ-to-germ lines or "blood-lines" concentrated in a given individual is fewer if the individual is inbred than if he is not. In other words, *the inbred individual possesses fewer different ancestors in some particular generation or generations than the maximum possible number for that generation or generations.*" Thus, according to the evolutionist's conception of the origin of species by natural selection, not only are all members of a species related in some degree, however remote, *but all members of all species from any one original life-spark presumably are members of one inbred line.* This wholly ridiculous conclusion follows, because the lines of descent terminating in any one individual, though they radiate back in widening angles for a time,

MATHEMATICAL CONSIDERATIONS 81

would be seen to gather together again in a comparatively few individuals if the pedigree of the species could be traced in its entirety.

Such a *reductio ad absurdum* is not altogether valueless. It shows how essential it is for one to recognize the unavoidable limitations, the desirability of definite analysis, the necessity of precise methods of attack, in any consideration of the proposition he may undertake.

There are three distinct phases of the inbreeding problem, as Pearl has pointed out:

1. The system of mating with regard to the relation of the actual number of ancestors making up the pedigree of an individual to the total possible number.

2. The constitution of each individual with respect to Mendelian unit factors which results from the continued operation of a given system of mating which is inbreeding.

3. The physiological effect produced upon the individual by the constitution derived from this system of mating.

The first two phases of the problem are capable of abstract mathematical treatment. The third can be solved only by experimental investigation.

Precise methods of measuring and comparing systems of mating have been devised by Pearl by the use of a Coefficient of Inbreeding and a Coefficient of Relationship.[a] The first is a measure of the actual number of

---

[a] Pearl has made a somewhat more precise analysis of the Inbreeding and Relationship Coefficients in later papers, 175, 176, and has suggested a Partial Inbreeding Index, in a percentage which one-half of the Relationship Coefficient is of the Inbreeding Coefficient. This constant is a measure of the amount of inbreeding due to relationship between the sire and dam. Further, he has described a single numerical measure of inbreeding for bisexual organisms, in the ratio of the area of the inbreeding curve in any pedigree to the area of the maximum (brother × sister) curve.

For our purposes, it is unnecessary to consider these extensions of Pearl's studies in detail, though technically they are very valuable.

ancestors compared with the possible number. It is derived from the formula:

$$Z_n = \frac{100(p_{n+1} - q_{n+1})}{p_{n+1}}$$

where p_{n+1} denotes the maximum *possible* number of different individuals involved in the matings of the $n+1$ generation, and q_{n+1} the *actual* number of different individuals involved in these matings. As an illustration, any individual in bisexual matings has two parents in the first ancestral generation, four grandparents in the second ancestral generation, and so on, according to the following symbolical representation

$$x \longleftrightarrow (1)\,2 \longleftrightarrow (2)\,4 \longleftrightarrow (3)\,8 \longleftrightarrow (4)\,16 \longleftrightarrow (5)\,32 \longleftrightarrow (n)\,2^n \ldots,$$

in which the enclosed numbers represent the ancestral generations (1 = parents, 2 = grandparents, 3 = great-grandparents, etc.), and the other figures the number of ancestors. In the second or earlier generations the ancestors may not all be different individuals, so that in any generation previous to the parental the actual number of ancestors may be less than the possible number. For example, in brother and sister mating, any individual instead of having four different grandparents, has only two. Expressed symbolically, as above, the representation for this type of mating would be

$$x \longleftrightarrow (1)\,2 \longleftrightarrow (2)\,4\text{-}y_1 \longleftrightarrow (3)\,8\text{-}y_2 \longleftrightarrow (4)\,16\text{-}y_3 \longleftrightarrow (5)\,32\text{-}y_4 \ldots,$$

where $y_1 = 2$, $y_2 = 6$, $y_3 = 14$, $y_4 = 30$.
In this case y has the value of $2^n - 2$, and this is the highest value it can have in any system of mating where two indi-

viduals are necessary for reproduction. Applying the formula given above, the Coefficients of Inbreeding for each generation in brother and sister mating are:

$$Z_0 = \frac{100\ (2-2)}{2} = 0$$

$$Z_1 = \frac{100\ (4-2)}{4} = 50$$

$$Z_2 = \frac{100\ (8-2)}{8} = 75$$

$$Z_3 = \frac{100\ (16-2)}{16} = 87.5$$

The figures obtained are the differences between the possible number of ancestors and the actual number expressed as percentages of the former. By plotting these percentages for successive generations on the generation number as a base, a curve of inbreeding is obtained which can be compared to the curves obtained by other systems of matings. This comparison is shown in Fig. 22 for the common types of matings as worked out by Pearl.

From these curves it is evident that continued brother by sister and double first-cousin matings have the same effect, although the latter is one generation behind the former. Also the curves for parent by offspring and single first-cousin matings are similar in type, but show the same differences in position. In any case the concentration of the lines of descent in these systems of inbreeding is rapid, until after fifteen generations no individual can have more than a fraction of one per cent. of the number of ancestors theoretically possible.

The Coefficient of Inbreeding alone tells us nothing as to the relation between the different lines of descent.

Two individuals may have the same Coefficients of Inbreeding when considered for any given number of generations, but differ greatly in germinal constitution. This is due to the fact that the two lines brought together in the immediate production of any individual may or may not be related. For example, a closely inbred animal of one breed may be mated to another closely inbred animal

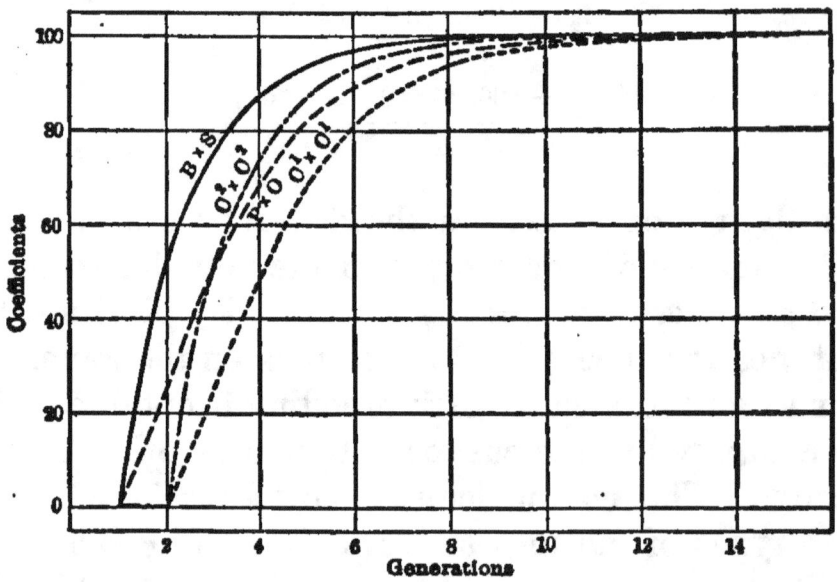

Fig. 22.—Curves of inbreeding showing (a) the limiting case of continued brother × sister breeding, wherein the successive coefficients of inbreeding have the maximum values; (b) continued parent offspring mating; (c) continued first cousin × first cousin mating where the cousinship is double ($C^2 \times C^2$), and (d) continued first cousin × first cousin mating where the cousinship is single ($C^1 \times C^1$). The continued mating of uncle × niece gives the same curve as $C^1 \times C^1$. (After Pearl.)

of an entirely different breed. The two lines of descent would then be totally unrelated as far as the known pedigrees are concerned, but the resulting individual would have a high Coefficient of Inbreeding, due to the concentration of ancestry separately in the two ancestral lines. To give some measure of the inter-relation of the lines of descent, Pearl has devised the Coefficient of Relationship, K, which is essentially the per cent. of the individuals in

MATHEMATICAL CONSIDERATIONS

each of the descending lines which are also represented in the other line. To give an adequate mathematical estimation of the degree of inbreeding, both constants are necessary. There is, generally, some correlation between them, although the Coefficient of Relationship may be zero, and the Coefficient of Inbreeding still be high, as in the illustration just given in which the progeny comes from a pair of individuals from two distinct inbred lines.

The application of these methods of determining the amount of inbreeding is illustrated by Pearl from the pedigrees of two Jersey bulls as follows:

Inbreeding "Z" and Relationship "(K)" Coefficients

of

King Melia Rioter 14th and Blossom's Glorene

A_0	$Z_0 (K_1)$	0	(0)	0	(0)
A_1	$Z_1 (K_2)$	25	(0)	0	(0)
A_2	$Z_2 (K_3)$	25.00	(50.00)	12.50	(0)
A_3	$Z_3 (K_4)$	37.50	(62.50)	12.50	(0)
A_4	$Z_4 (K_5)$	50.00	(75.00)	25.00	(0)
A_5	$Z_5 (K_6)$	71.88	(87.50)	29.69	(0)
A_6	$Z_6 (K_7)$	81.25	(92.19)	35.94	(0)
A_7	$Z_7 (K_8)$	90.63	(92.97)	40.23	(0)

The method of making the calculations is explained clearly and concisely by the originator and we shall not undertake to repeat it here. What we are interested in is the genetic meaning of the figures after they have been obtained.

The Coefficient of Inbreeding, Z, has to do solely with total relationship, and shows the intensity of inbreeding in the stockman's sense of the word by measuring precisely "the proportionate degree to which the actually existent number of different ancestral individuals fails to reach the possible number, and by specifying the location in the series of the generation under discussion." King Melia Rioter 14th had less than 10 per cent. of the maximum number of ancestors in the 7th ancestral generation, while in the same generation Blossom's Glorene had nearly 60 per cent. From these figures it is evident that King Melia Rioter 14th is a much more inbred animal than

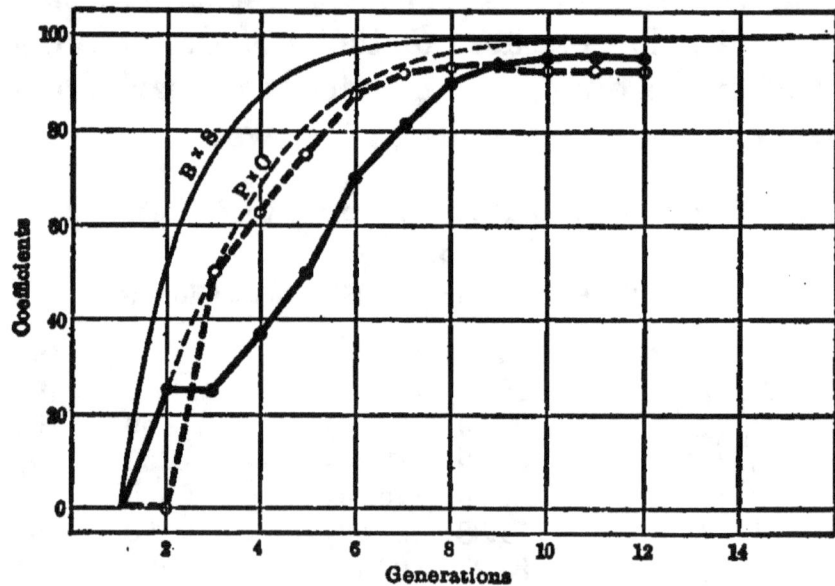

Fig. 23.—Graphs showing (a) the total inbreeding (heavy solid line) and (b) the relationship (heavy broken line) curves for the Jersey Bull, King Melia Rioter 14th. The high order of the inbreeding and relationship between the sire and dam in this case is evident by comparison with the lighter lines, which give the maximum values for continued brother × sister and parent × offspring breeding. (After Pearl.)

Blossom's Glorene. A clearer demonstration of the matter, however, is found in Fig. 23, where the curve of the total inbreeding of King Melia Rioter 14th plotted from the figures just cited is compared with the curve of maximum values for continued brother × sister and parent × offspring matings.

The Coefficient of Relationship, K, might better be called the Coefficient of Cross-Relationship to distinguish its function from that of the Coefficient of Inbreeding, since it is a measure of the community of ancestry of the dam and the sire.

These two coefficients taken together, then, give us the first quantitative measure of inbreeding as a system of mating, but obviously they do not tell anything concerning the actual germinal constitution of any individual resulting from a given system of inbreeding. This feature of the relationship coefficients is nicely illustrated by one of Pearl's examples. Clearly, a Holstein cow produced by continued brother × sister matings (K = 100) is very different in its germinal constitution from a crossbred animal obtained by mating this cow with a Jersey bull, the product of a similar system of inbreeding ($K = 0$). Yet the Coefficients of Inbreeding in each case form identical series, with the maximum possible value of Z when $K=0$ one generation farther removed than when $K=100$.

Without question the germinal (or may one call it the Mendelian?) composition of any individual can be determined only by actually testing its breeding qualities, its transmissive powers; and the effect this composition may have had upon its development can be measured only by comparison with other individuals of known genetic constitution. But an indication of the germinal constitution of an individual produced by any long-continued system of inbreeding, as far as the degree of heterozygosity or homozygosity is concerned, can be obtained by applying the laws of probability to Mendelian formulæ. In other words, the laws of probability applied to Mendelian formulæ show the probable homozygosity or heterozygosity of the generation as a whole for any number of Mendelian allelomorphic pairs with any given system of inbreeding, and some idea of the composition of the individual may often be had from a careful consideration of the composition of the generation to which it belongs.

In an endeavor to demonstrate the effect of various systems of inbreeding upon Mendelian constitution, and to appraise the effect of this constitution upon developmental vigor, let us approach the problem from the opposite direction.

It has been established that the effect produced by crossing depends more or less closely upon the genetic diversity of the types which produce the hybrid. The usual result of crossing organisms which differ in many characters is a first generation which is no more variable than the parental types. The second generation, however, may be expected to show a greater variability because of Mendelian segregation. The amount of such variability is a measure of the diversity of the parents which produce the cross. It is in crosses which show greater variability in the second generation that hybrid vigor is expected in the generation immediately following the cross. When such hereditary combinations are composed of unlike elements, hybrid vigor is commonly shown; when all the combinations are composed of like elements hybrid vigor is absent. Hence, in crossed species of wild or domesticated animals and plants part of their vigor may be the result of dissimilar hereditary factors acting together. If conditions are brought about by which this dissimilarity in allelomorphic combinations is reduced or lost completely a partial diminution of developmental energy will occur. Since there is a constant tendency for inbreeding of whatever kind to bring about similarity in germinal construction, inbreeding will,

MATHEMATICAL CONSIDERATIONS

therefore, frequently cause a general reduction in vigor. It has never been held that all hereditary factors are equally involved in this effect on vigor. Some are considered to be wholly without effect. The fact remains, however, that the increased growth and extra vigor commonly resulting from hybridization as a mass effect is intimately associated with Mendelian phenomena, and its expression is roughly proportional to the number of heterozygous factors present and disappears when homozygosity is brought about.

The reduction of the number of heterozygous allelomorphs in an inbred population is automatic and varies with the closeness of inbreeding. In self-fertilization it follows the well-known Mendelian formula by which any heterozygous pair forms in the next generation 50 per cent. homozygotes and 50 per cent. heterozygotes in respect to that pair. Since the homozygous allelomorphs must always remain constant and the number of heterozygous factor combinations is halved each generation and one-half added to the homozygous class, the reduction in the number of heterozygous elements proceeds as a variable approaching a limit by one-half the difference in each generation. The curve illustrating this approach to complete homozygosity is shown as No. 1 in Fig. 24.

As East and Hayes[59] have said: "Mendel, in his original paper, showed that if equal fertility of all plants in all generations is assumed, and, furthermore, if every plant is always self-fertilized, then in the nth generation the ratio of any particular allelomorphic pair (A, a) would be $2^n - 1 AA : 2Aa : 2^n - 1 aa$. If we consider only homozygotes and heterozygotes, the ratio is $2^n - 1 : 1$. Of course, the matter is not quite so simple when several allelomorphs are concerned, but in the end the result is similar. Heterozygotes are eliminated and homozygotes

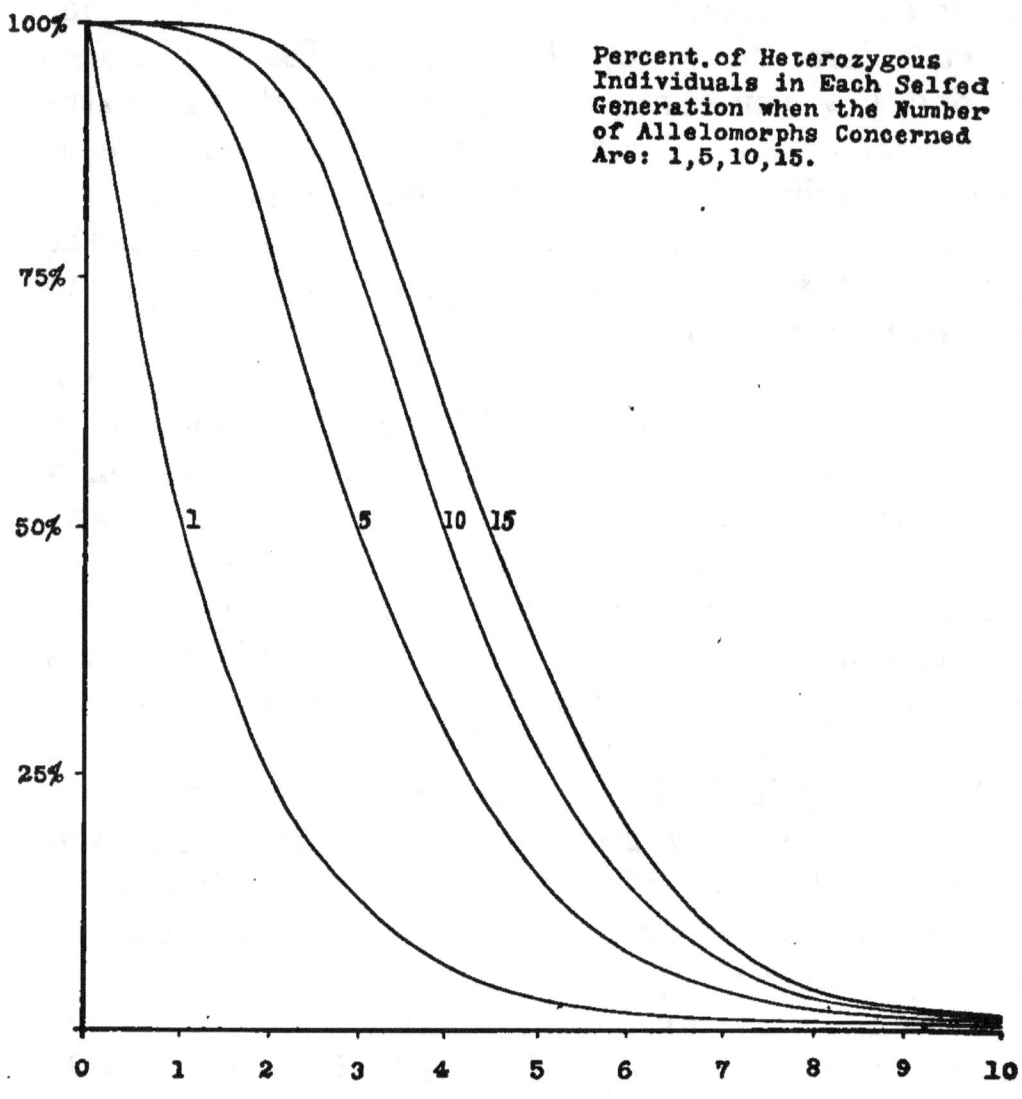

FIG. 24.—Graphs showing the reduction of heterozygous individuals and of heterozygous allelomorphic pairs in successive generations of self-fertilization.

remain. The probable number of homozygotes and any particular class of heterozygotes in any generation r is found by expanding the binomial $1 + (2^r - 1)^n$ where n represents the number of character pairs involved. The exponent of the first term gives the number of heterozygous and the exponent of the second term the number of homozygous characters. As an example, suppose we desire to know the probable character of the fifth segregating generation (F_6) when inbred, if three character pairs are concerned. Expanded we get

$$1^3 + 3 [1^2 (31)] + 3 [1 (31)^2] + (31)^3$$

Reducing, we have a probable fifth-generation population consisting of 1 heterozygous for three pairs; 93 heterozygous for two pairs; 2883 heterozygous for one pair; 28,791 homozygous in all three character combinations." Of the 32,768 total number of individuals in this generation, 2977, or 9.09 per cent., are heterozygous in respect to some characters. Of the 98,304 total number of allelomorphic pairs involved in all the individuals of this generation, 3072, or 3.125 per cent., are heterozygous. This is the percentage which is obtained by halving 100 per cent. five times. It is the per cent. of heterozygous allelomorphic pairs in all the individuals making up the population as a whole that follows curve 1 in Fig. 24. The per cent. of individuals heterozygous in any factors in any generation inbred by self-fertilization depends upon the number of heterozygous elements concerned at the start. The curves where 1, 5, 10 and 15 heterozygous allelomorphs are present in the beginning are given in Fig. 24. These are calculated from the formula given and illustrated above. The curve for the reduction in heterozygous individuals where one factor only is concerned at the start, is identical with the curve showing the reduction in the number of heterozygous factors in an inbred population as a whole where any number of factors are concerned. In any case, almost complete homozygosity is reached in about the tenth generation.[b]

It must be remembered that this reduction applies only to the whole population, or to a representative sample of the population, in which every member is selfed, in which each individual is equally fertile, and in which all the progeny are grown in every generation. In practice in an inbreeding experiment, usually only one individual in self-fertilization or two individuals in brother and sister matings are used to produce the next generation. Thus the rate at which complete homozygosity is approached depends on the constitution of the individuals chosen. Theoretically in any inbred generation the progenitors of the next generation may either be completely heterozygous or completely homozygous or any degree in between depending upon chance. The only conditions which must follow in self-fertilization is that no individual can ever be more heterozygous than its parent, but may be the same or less. Thus it is seen that artificial inbreeding, as it is practiced, may theoretically never cause any reduction in heterozygosity, or it may bring about complete homozygosity in the first inbred generation. In other words, the rate at which homozygosity is approached may vary greatly in different lines. However, as the number of heterozygous factors at the commencement of inbreeding increases the more nearly will the reduction to homozygosity follow the curve shown, because the chance of choosing a completely

---

[b] Various formulæ dealing with inbreeding have been proposed and discussed by Pearson (177), Jennings (102, 104, 105, 106,) Pearl (168, 169, 170, 171, 172), Fish (69), Wentworth and Remick (213), Robbins (186, 187) which are useful in predicting the character of inbred generations when certain conditions are fulfilled.

homozygous or completely heterozygous individual in the early generations will become very much less.

TABLE II

THE THEORETICAL NUMBER AND RATIO OF INDIVIDUALS IN THE CLASSES OF DIFFERENT DEGREES OF HETEROZYGOSITY, AFTER RECOMBINATION, WHEN FIFTEEN MENDELIZING UNITS ARE INVOLVED.

Class No.	The total number of individuals in all the possible Mendelian recombinations in F_2 when 15 factors are involved.	Ratio of individuals in the classes with different number of heterozygous and homozygous factors — coefficients $(a+a)$.[15]	The number of factors in respect to which the different classes are:		The total number of heterozygous and homozygous factor pairs in all the individuals in each class:	
			Heterozygous	Homozygous	Heterozygous factor pairs	Homozygous factor pairs
1	32,768	1	15	0	15	0
2	491,520	15	14	1	210	15
3	3,440,640	105	13	2	1,365	210
4	14,909,440	455	12	3	5,460	1,365
5	44,728,320	1,365	11	4	15,015	5,460
6	98,402,304	3,003	10	5	30,030	15,015
7	164,003,840	5,005	9	6	45,045	30,030
8	210,862,080	6,435	8	7	51,480	45,045
9	210,862,080	6,435	7	8	45,045	51,480
10	164,003,840	5,005	6	9	30,030	45,045
11	98,402,304	3,003	5	10	15,015	30,030
12	44,728,320	1,365	4	11	5,460	15,015
13	14,909,440	455	3	12	1,365	5,460
14	3,440,640	105	2	13	210	1,365
15	491,520	15	1	14	15	210
16	32,768	1	0	15	0	15
16	1,073,741,824	32,768	15	15	245,760	245,760
$n+1$	$(2n)^2$	$2n$	n	n	$\frac{1}{2}(n \cdot 2n)$	$\frac{1}{2}(n \cdot 2n)$

As an example, in Table II there is shown the theoretical classification of the progeny of a self-fertilized organism which is assumed to be heterozygous with respect to 15 independent Mendelizing units. It can be seen that the bulk of the individuals lie between classes 6 and 11, where none of the members is heterozygous for more than 10 or less than 5 factors. In other words, any individual selected from this population to be the progenitor of the next generation would most probably come from the middle classes and, therefore, would be heterozygous for only about half as many factors as its parent. The chance that this individual would not come from the middle classes between 6 and 11 would be about 1 out of 10. The chance that it would be completely homozygous or completely heterozygous is 1 out of 32,768. If 20, instead of 15, factors were involved, the chance would be 1 out of 1,048,576. The selection of such completely homozygous individuals would be a remarkable event. If, for instance, a tobacco plant, which has 24 chromosomes as the haploid number, could be obtained which was heterozygous in one factor pair in each chromosome and this plant were to be self-pollinated and the progeny grown, 16,777,216 plants would have to be produced in order to provide an even chance of securing somewhere in the lot one plant which was homozygous in all the twenty-four factors. This number of plants would require over 2000 acres of land as tobacco is grown in field culture.

This condition by which the progenitor of each generation in self-fertilization tends to be half as heterozygous as its parent holds true for any number of factors and in every generation. Thus it can be seen (Table II) that the progeny as a whole have an equal number of heterozygous and of homozygous factor pairs in respect to those characters in which the parent was heterozygous. So it is that in practice the reduction in heterozygosity accompanying inbreeding is greatest at first, rapidly becomes less and finally ceases for all practical purposes. From 0 at the start the degree of homozygosity, with respect to a given number of factors, increases to 99 per cent. after 7 generations of self-fertilization; after 12

MATHEMATICAL CONSIDERATIONS

generations it is 99.9 per cent. and after 19 generations 99.999 per cent.

Although nearly complete homozygosis is theoretically brought about by 7 generations of self-fertilization the attainment of absolute homozygosity is a difficult matter and in practice it may never be reached. The fact that hereditary factors are distributed by the chromosomes, so that there is not independent recombination among all the determiners, enters as a complicating factor. Lethal factors, which prevent homozygotes from appearing, and increased productivity of hybrid combinations, also tend to prevent the complete elimination of heterozygosity.

The way in which factor linkage affects reduction to homozygosity may be illustrated by the use of two allelomorphic pairs of factors. Jennings [106] has calculated the effect of three generations of self-fertilization upon the population descending from a dihybrid when the two pairs of factors show a linkage relation of 2 (that is, 33⅓ per cent. "crossovers") in both sexes; and also when the linkage is complete in one sex, as in Drosophila where there is no "crossing over" in the male. The proportions of completely homozygous, of completely heterozygous, and of mixed individuals—i.e., heterozygous in one pair and homozygous in the other—obtained after three generations of self-fertilization, are compared with what is expected when the two factors are independent as follows:

Ratio required	Two factors independent	Linkage ratio 2 in both sexes	Linkage complete in one sex, 2 in other
Per cent. of complete homozygotes..	76.56	77.14	78.70
Per cent. of complete heterozygotes.	1.56	2.14	3.70
Per cent. homozygous in one pair but not in other	21.88	20.71	17.59
Per cent. homozygous factors.......	87.50	87.50	87.50
Per cent. heterozygous factors	12.50	12.50	12.50

It will be seen from these figures that the proportion of complete homozygotes and complete heterozygotes is increased by linkage at the expense of the mixed class. The proportion either of homozygous or of heterozygous factor pairs, however, is unaffected. It is evident then that just as the reduction to homozygosity by self-fertilization is independent of the number of factors involved, in the same way it is independent of the way in which these factors are linked together: but in an experiment where particular individuals are chosen as progenitors linkage of factors reduces the chance that these will come from the median classes of heterozygosity; hence, the rate at which homozygosity is attained will vary more widely between dfferent lines if the factors involved are partially linked than if they are all independent. This merely means that some lines will become uniform and lose the stimulus of hybridization in a fewer number of generations than will other lines and that this difference theoretically is increased by linkage. But the hastening of the attainment of homozygosity in some lines is balanced by delay in other lines, so that on the average the curve of inbreeding shown applies equally whether linkage of factors is involved or not.

If there were no other controlling factors the reduction in vigor resulting from inbreeding, in the majority of cases, should approximate curve 1 in Fig. 24 on the assumption that hybrid vigor or heterosis is associated with heterozygosity. However, it should not be thought that the amount of heterosis is perfectly correlated with the number of heterozygous factors. Some have more of an effect than others, and certain factors, when combined together, may have a cumulative effect. Moreover, since the heterozygous individuals are more vigorous than the homozygous, selection either unconscious or purposeful would favor the more heterozygous so that actual approach to homozygosity is quite likely not to proceed at as fast a rate as the theoretical curve would indicate.

Self-fertilization is the quickest and surest means of obtaining complete homozygosity for the reason that whenever any pair of allelomorphs becomes homozygous it must always remain so, as long as self-fertilization takes place, whereas in brother and sister mating a homozygote may be mated to a heterozygote. The approach to homozygosity in self-fertilization when one pair of contrasted characters is considered and fecundity does not vary proceeds as follows:

Generation	F_1	F_2	F_3	F_4		
Parent type AA	0	1/4	3/8	7/16	..............1/2	$AA_6 = .492$
Parent type aa	0	1/4	3/8	7/16	..............1/2	$aa_6 = .492$
Hybrid type Aa	1	1/2	1/4	1/8	 0	$Aa_7 = .008$

In brother and sister mating the procedure is as follows:

Generation	F_1	F_2	F_3	F_4	F_5		
Parent type AA	0	1/4	2/8	5/16	11/32	.....1/2	$AA_{17} = .490+$
Parent type aa	0	1/4	2/8	5/16	11/32	.....1/2	$aa_{17} = .490+$
Hybrid type Aa	1	1/2	2/4	3/8	5/16	0	$Aa_{21} = .008+$

These figures have definite numerical relations to each other and formulæ have been obtained by Jennings [105] for calculating the condition in any generation. It will be seen from the above figures that 6 generations of self-fertilization are more effective than 17 generations of brother and sister matings in bringing about homozygosis.

Of parent by offspring matings there are several

methods which may be carried on with different results. If the individuals are mated at random to all the different types of parents, homozygous and heterozygous, the effect is the same as in brother and sister mating. Cousin matings, which may proceed either as single or double first-cousin matings or even more distant unions, may be equally or less effective; but all tend towards the same end; heterozygosity is ultimately eliminated and homozygosity prevails. In this way there is a difference between selective mating and random mating. Continued selective mating is necessary to bring about homozygosity. Intermittent inbreeding alternating with periods of outcrossing which is the prevailing state of affairs with many organisms cannot maintain any high degree of homozygosity.

In self-fertilization the reduction in heterozygous allelomorphs in a population as a whole follows curve 1 in Fig. 24, irrespective of the number of factors concerned, as stated before, provided that a random sample of all the different classes of individuals are selfed and become progenitors of the next generation and that there is equal productiveness and equal viability. If the heterozygotes are more productive, as in many cases they are, the reduction to complete homozygosity will be delayed.

Artificial self-fertilization in naturally crossed species, then, brings about the same condition as prevails in naturally selfed species. The great variability of a cross-fertilized species gives way to the more uniform and stable condition characteristic of naturally self-fertilized organisms. The uniformity brought about by inbreeding is thus due to a reduction of the genetical variability. Inbreeding affects physiological or developmental vari-

MATHEMATICAL CONSIDERATIONS

ability only indirectly by changing the vigor of the organisms so that they may react differently to different environments.

Assuming, then, that the loss of the stimulation accompanying heterozygosity is correlated with the reduction in the number of heterozygous factors we should expect to find the decrease of heterosis greatest in the first generations, rapidly becoming less until no further loss is noticeable in any number of subsequent generations of self-fertilization, and that on the average the decrease will become negligible from the seventh to the twelfth generation and from then on no further marked change will take place. Segregation of characters and appearance of new types and reduction in variability will also follow the same course. Some cases are to be expected in which stability is reached earlier, and some cases in which it is reached later; or, theoretically it may never be reached. With these points in mind, let us see what are the actual results of long-continued inbreeding.

CHAPTER VI

INBREEDING EXPERIMENTS WITH ANIMALS AND PLANTS

Doubtless discussion has been rife since the dawn of civilization as to the actual effect of more or less close intermating in the various breeds of domestic animals, since stock-raising was one of the earliest arts and was brought to a high degree of perfection by the ancient Semitic nations. One may surmise, from the rules they made against the marriage of near relatives, that the proponents of cross breeding had the best of the argument; but it is hardly likely that their practice was anything more than rule-of-thumb adopted after a variety of casual observations. At any rate, controversy is still spirited, and one reads the opinion of stock-breeding authorities without arriving at any definite knowledge of the problems. Their results are confusing, and the only conclusion one may reach from their perusal is the wholly unsatisfactory one that close mating, as a system of breeding, has both advantages and disadvantages. Without question, it has had great value in fixing certain desirable types. Some breeds, as a whole, and many individual herds, owe their uniformity in conformation and performance in a large measure to close inbreeding accompanied by rigid selection. At the same time, it must be recognized that certain evil effects may result from close intermating. These effects have been frequently expressed in lessened constitutional vigor, greater susceptibility to disease, reduced fecundity, and, in some cases, even in decreased size and in the appearance of definite abnormal or pathological conditions.

Obviously it is not possible to formulate any definite rule by which to judge under what condition the good effects of inbreeding may be expected to outweigh the evil, by generalizing from a series of isolated facts. What is needed is controlled experimentation to determine just what inbreeding involves, and interpretation of the results in keeping with general biological knowledge. Darwin[38, 39] was the first to appreciate this. After endeavoring rather unsuccessfully to generalize on the subject with a collection of published records as a basis, he himself carried on a series of inbreeding experiments on plants extending over a period of eleven years. Plants were probably selected as the material with which to work primarily because the experiments could be carried out easily and with little expense. But there is another reason why plants serve the purpose better than animals in such an investigation; the animals commonly available are bisexual; hence, they cannot be inbred as intensively as hermaphroditic plants. Nevertheless, Darwin's experiments served to stimulate more interest in the subject among zoölogists than among botanists, and the quantitative experiments carried on during the next thirty years were nearly all upon animals.

The rat has been a favorite species, it having served as material for the extended researches of Crampe,[34] Ritzema-Bos[184] and King.[119, 120, 121] The first two investigations, together with that of Weismann and von Guaita,[86, 87] on mice have been the classic examples of the adverse effects of inbreeding.

Crampe's experiments started with a litter of five

young, obtained by crossing an albino female with a white and gray male. These animals were inbred in various degrees for seventeen generations. During the experiment many rats showed great susceptibility to disease, divers kinds of abnormalities, diminished fertility, and increased total sterility.

Similarly Ritzema-Bos started his investigations with a litter of twelve rats obtained by crossing, this time an albino female with a wild Norway male. This stock was inbred in different ways for six years, during which time he claimed to have obtained about thirty generations. His results did not corroborate those of Crampe in so far as susceptibility to disease or appearance of malformations are concerned, but there was a gradual decrease in size of litter and a gradual increase in percentage of infertile matings, as is shown in the following table:

Year of inbreeding	1	2	3	4	5	6
Ave. number in litter	7.5	7.1	7.1	6.5	4.2	3.2
Per cent. infertile matings	0.0	2.6	5.6	17.4	50.0	41.2

These investigations, in spite of the habit biologists have of citing them, are not calculated to settle the question they undertook to answer. Ritzema-Bos himself criticizes those of Crampe, because he believes them to have been started with a weak strain. Miss King, however, thinks the weakness of these rats, as indicated by their susceptibility to disease, the appearance of malformations, and their tendency to sterility, was due to the conditions under which they were kept. She had a similar experience during the earlier part of her own experiments, and found that inadequate nourishment was largely the cause. But it is not for this reason that we feel that both of these experiments should be disregarded. *Each was started with hybrid stock, and such experiments with hybrid stock bring in an additional complication, Mendelian recombination.* The only type of investigation on bisexual animals calculated to offer critical evidence on the effect of inbreeding *per se* must be carried on with stock which has already been inbred long enough to reduce the genetic constitution of the animals to an approximately homozygous condition. Then, and then only, can the effect of more extended inbreeding be determined without confusion as to the interpretation of the results.

Miss King has pointed out a part of the difficulties involved in starting with a hybrid stock. In one of her experiments the progeny of a cross between a wild Norway rat and an albino was inbred for several generations. She found that while the majority of the F_1 females were fertile, at least 25 per cent. of the F_2 females were completely sterile and 10 per cent. of those which did breed cast only one or two litters. In the strains extracted from this cross there was variation in degree of fertility, but none was found which exhibited the high degree of fertility usually existent in the albino rat. No endeavor to select fertile strains was made and one cannot say whether or not rigid selection would have isolated them, but the researches of Detlefsen [47] on hybrids of the genus Cavia to which the common guinea-pig belongs indicate this to be a probability. These investigations as well as those of East [53] on the genus Nicotiana show conclusively that various hereditary factors are involved in the partial sterility exhibited in many species crosses, and that these factors may be expected to recombine in the usual manner.

It is more than a mere assumption, then, if a great part of the sterility found by Crampe and Ritzema-Bos is attributed to the same cause.

The investigations of Weismann and von Guaita are hardly more satisfactory. Weismann inbred a stock of white mice for twenty-nine generations, and found the average number of young for the three 10-year periods to be 6.1, 5.6 and 4.2. Where this stock originated, and what method of inbreeding was followed, we are not told. Presumably the gross result was a slight decrease in fertility, coincident with the amount of inbreeding, but even this is not certain. As King points out, the average number of litters under observation in the first two generations was twenty-two; in the last nine generations, three. Clearly there was greater opportunity of selecting healthy breeding stock, as well as a lower probable error, in the earlier part of the experiment, and this might account for the slight difference in fertility found. Von Guaita crossed some of these highly inbred mice with Japanese waltzers and then inbred for six generations. He reports the average number of young in the successive generations as 4.4, 3.0, 3.8, 4.3, 3.2 and 2.3; but in view of the vigor almost always expressed in the F_1 generation of such crosses one is inclined to doubt the pertinence of these figures to the inbreeding problem.

Although, as has been pointed out, there is good reason for disallowing the claim of these much cited experiments on mammals as proofs of the adverse effects of inbreeding through consanguinity alone, there is no intention of denying the isolation of individuals characterized by undesirable qualities from mixed strains by means of Mendelian recombination. Perhaps it is not wise even to maintain the impossibility of injury to *any strain of any species* through inbreeding *per se*, but it is proper to say that the evidence in favor of it is practically *nil*.

Doubtless we could make our case more convincing to the stockman could the enormous number of really well-kept herd records be cited and analyzed. But it is not possible at present to say whether many of these records satisfy the requirements of modern genetic research. This is a task which must be left to the breeding organizations of the future. We can appeal at present to only two investigations on mammals where the effect of Mendelian recombination has been largely eliminated; and these again are on small mammals, the rat and the guinea-pig.

The first of these investigations to be reported was that of King.[119, 120, 121] It was started with a litter of four slightly undersized but otherwise normal albino Norway rats, two males and two females. From these females two lines, *A* and *B*, were carried on for twenty-five generations by mating brother and sister. In the earlier generations practically all of the females were used for breeding, but in every generation after the sixth about twenty females were selected from approximately a thousand available young.

At first the inbred rats exhibited many of the defects reported by Crampe. Numerous females were either sterile or produced but one or two small litters. Other animals were characterized by low vitality, dwarfing, and malformations. Stock rats exhibiting the same characteristics at this time, however, led to a change in the food, following which the "dire effects of inbreeding" practically disappeared. Whether this improvement in the colony was due entirely to the change of diet or may be attributed partly to selective elimination of the weaker rats cannot be determined. We are inclined to agree with Miss King in giving greater weight to the first factor, though for a reason which she does not mention. The general success of Miss King's whole investigation we believe to be due largely to the fact that the experiments were started with stock rats which already must have been very closely inbred and therefore in an approximately homozygous condition.

From the seventh generation on, selection was made on the new-born young with general vigor as the basis, but the two lines were selected differently. In line A only litters having an excess of males were selected to serve as the progenitors of the succeeding generation, while in line B the reverse was the case. The general result was to show that the normal sex ratio in this species, 105 males to 100 females, can be changed. At the end of nineteen generations of selection, line A had produced litters having a sex ratio of 122.3 males to 100 females, and line B had produced litters having a sex ratio of 81.8 males to 100 females. From these facts there is no doubt but that lines having an hereditary tendency to produce different sex ratios can be isolated, but there is no evidence whatever in favor of the theory of Düsing proposed in 1883 to the effect that inbreeding by lessening the vitality of the mother increases the percentage of male young. The change in the sex ratio was made in two generations. After that the effect of selection ceased. Such a result not only militates against attributing the changed ratios to inbreeding itself, but indicates that a relatively small number of Mendelian factors are involved in the control.

The effect of continuous inbreeding on body weight is

shown in Fig. 25. This graph is constructed from data collected from the records of males of line *A*, but graphs constructed from the records of the females of this line and from males and females of line *B* do not differ from

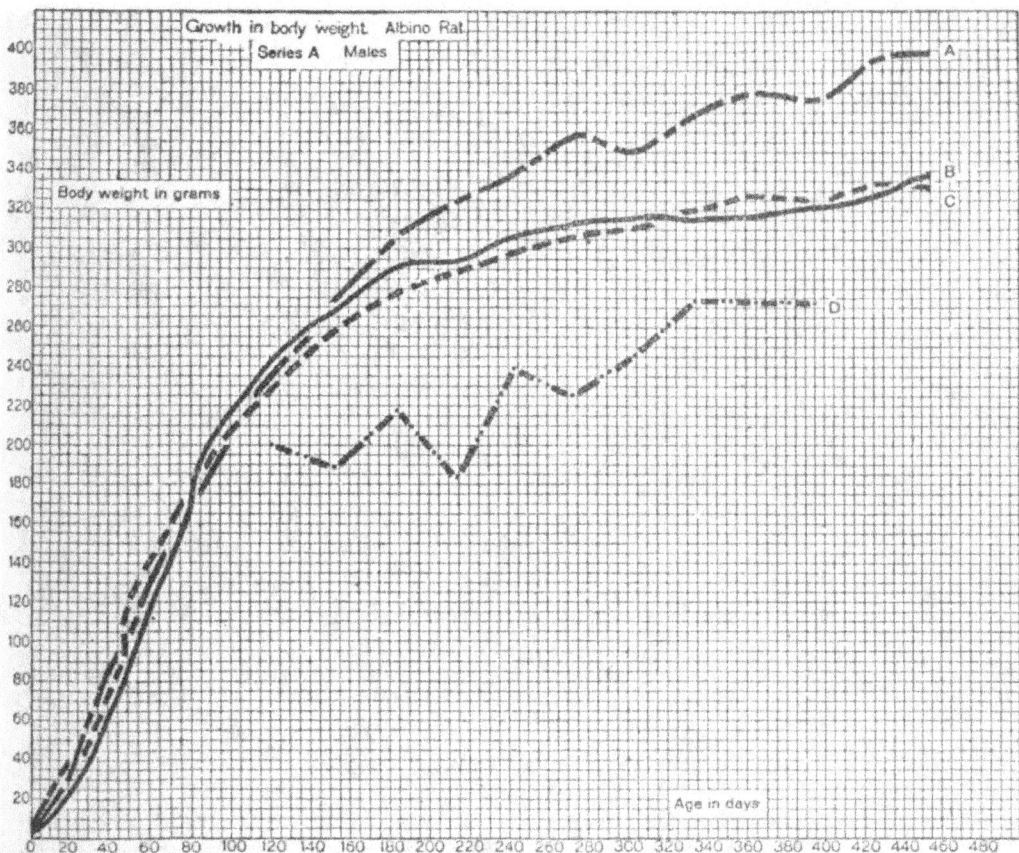

Fig. 25.—Graphs showing the increase in the body weight with age for males of inbred albino rats. (Series A.) A, graph for the males of the seventh to the ninth generations inclusive; B, graph for the males of the tenth to the twelfth generations inclusive; C, graph for the males of the thirteenth to the fifteenth generations inclusive; D, graph for the males of the first six inbred generations. (After King.)

it in any essential feature. Curve *D* is further evidence for concluding that the animals of the first six generations suffered from malnutrition, since, as Miss King notes, it is preposterous to suppose that these animals could have given rise to the very large individuals represented by curve *A* if it really represented their true body weight. How favorably these inbred strains compare with stock animals is shown in Fig. 26.

Paralleling the results obtained for body weight were

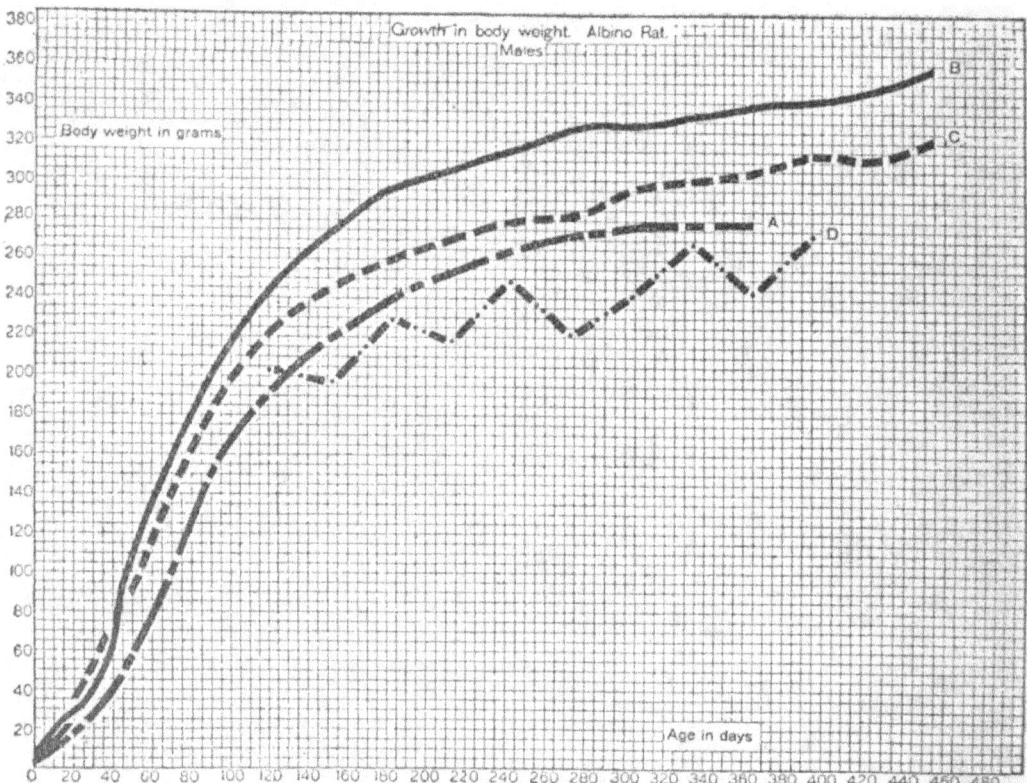

Fig. 26.—Graphs showing the increase in the weight of the body with age for different series of male albino rats. A, graph constructed from Donaldson's data for stock albinos; B, graph for males belonging in the seventh to the fifteenth generations of the two series of inbreds combined; C, graph constructed from data for a selected series of stock albinos used as controls for the inbred strain; D, graph for males belonging in the first six generations of the two series combined. (After King.)

those upon fertility and constitutional vigor as judged by longevity. Neither was reduced by inbreeding; in fact, there seems to be no doubt but that there was a significant increase in both cases. There was a slight but definite increase in fertility as is evident if one plots the theoretical curve which fits the experimental curve for litter size throughout the course of the experiment (Fig. 27). The whole series of inbreds compares well in this respect with stock albinos for which the average litter size is 6.7. Further there was a notable increase in longevity in line *A* and a marked increase in line *B*.

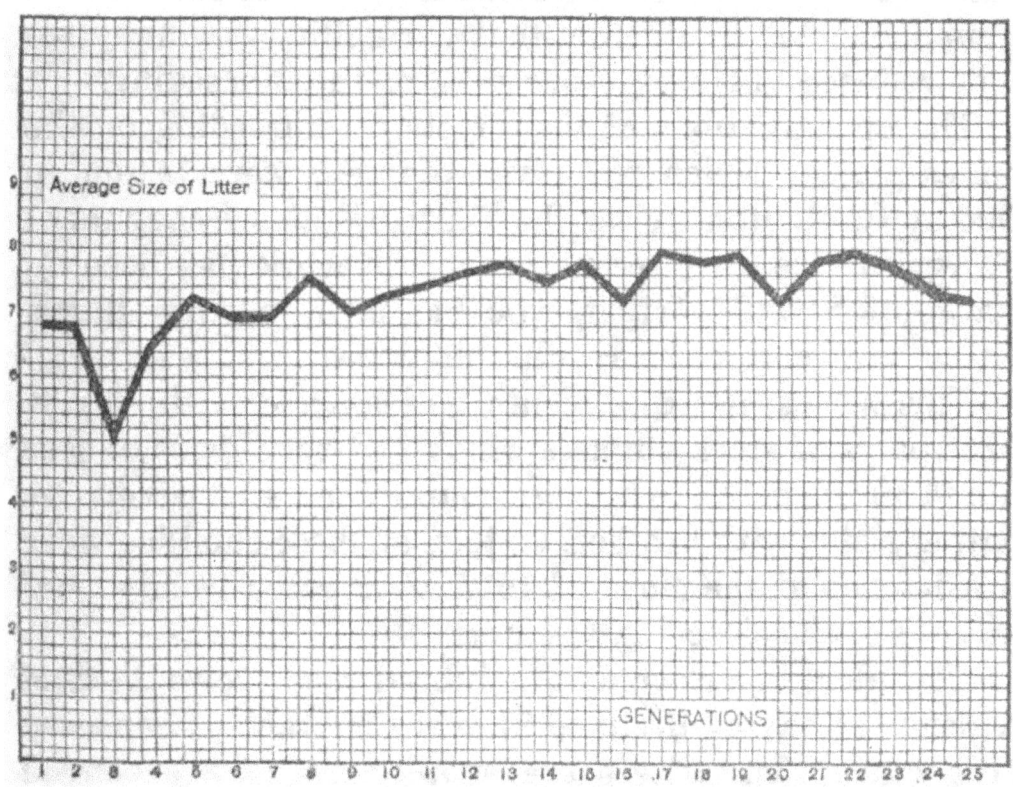

Fig. 27.—Graph showing the average size of litters produced in successive generations of inbreeding albino rats by brother and sister matings. (After King.)

The interpretation of these experiments is wholly in accordance with Mendelian theory. Starting with stock rats which from previous close breeding had already been reduced to a high degree of homozygosity, inbreeding had the tendency to accentuate this purity of type and to segregate slight differences. By selection vigorous uniform strains were built up, strains somewhat larger, more fertile and longer lived than many strains of stock rats. It is clear that this was the result of Mendelian recombination for the two lines A and B were in the end somewhat different. The rats of line A were slightly more fertile, attained sexual maturity earlier, and lived longer than those of line B. If this evidence were not sufficient, it is supplemented by the fact that variability gradually became reduced during the progress of the inbreeding.

The investigations of the effects of inbreeding on the guinea-pig, to which we have referred, were begun in 1906 by G. M. Rommel of the United States Department of Agriculture. In recent years the work has been in charge of Sewall Wright, who has made a very illuminating analysis of the results obtained.

This series of experiments was started with thirty-three pairs of stock animals which had been more or less inbred previously. Although maintained exclusively by mating sister with brother, sixteen of these families were in existence at the close of 1917 after some twenty generations of the closest inbreeding.

Considered as a whole this inbred race shows distinct evidence of having declined in every character connected with vigor. The litters are smaller and are produced more irregularly. The per cent. of mortality both *in utero* and between birth and weaning has increased. The birth weights are lower and the rate of growth slower than in control stock. In spite of these facts, however, one is forced to the conclusion that these results are not the effect of inbreeding as a direct cause, but are to be attributed to Mendelian segregation.

There are pronounced differences between the various families. Some are still very vigorous, comparing favorably with the original stock; others degenerated so rapidly that they soon became extinct in spite of every effort to prevent such a catastrophe. Among the families still in existence, there is even evidence that vigor as a general term may be divided into various causative factors and that these factors may be combined in various ways. By grading each family for various characters connected with vigor of growth and reproduction and then classifying each family in numerical order for each separate character, Wright has been able to show conclusively that there are many hereditary factors which affect fertility, growth and vitality and that almost any combination of these characters may become fixed in a family through inbreeding.

A little later we shall have occasion to speak of several noteworthy end results obtained by inbreeding the larger domestic mammals, but no further discussion seems advisable in this place because of the lack of quantitative data. A similar statement holds for birds.

The fruit fly, *Drosophila melanogaster*, is the only insect which has been used for extended experiments on effects of inbreeding, although there are numerous examples on record where an importation of a relatively small number of individuals has resulted in an overwhelming increase—witness the gypsy moth in New England.

Castle and his co-workers [21] bred Drosophila for many generations by continuous brother and sister matings. After fifty-nine generations of this close inbreeding the fertility did not appear to be reduced below that shown by the original stock, although it was increased by crosses between certain inbred lines. There was some indication of reduction in size when inbred flies were compared with random mated stock reared under the same conditions. Far from being exterminated by inbreeding, however, the flies at the end of the experiment were apparently fully equal to those with which it was begun.

These experiments showed clearly that inbreeding results in strains of unequal fertility. The less fertile tended to be eliminated by differential productiveness, so that only the more fertile remained. The occurrence of absolute sterility was pronounced in the first part of the experiment, but almost entirely disappeared in the later generations. The figures as calculated from their table are as follows:

	Per cent. of matings totally sterile
Generations 6 to 24	17.80
Generations 25 to 42	18.47
Generations 43 to 59	3.37

Such a result is to be expected when it is remembered that inbreeding produces homozygous individuals, and these, whenever sterile, are, of course, eliminated.

Moenkhaus,[144] Hyde,[98] and likewise Wentworth,[211] by similar inbreeding experiments with Drosophila found sterility, though increased in the first stages of inbreeding, tended to be eliminated after the process was long continued.

The only other experiments on invertebrates which ought to be cited here are those of Whitney[216] and A. F. Shull[198] on the rotifer *Hydatina senta*. Both of these investigators found that inbreeding had a considerable adverse effect on the size of family, number of eggs laid per day, rate of growth, and variability. The proper interpretation of their results is somewhat obscure, unless one hypothecates the origin of frequent mutations. The number of generations bred and the number of families under observation were not sufficient to demonstrate the segregation of differences in these characteristics, though this is to be expected since these qualities are symptomatic of general vigor and general vigor was increased by crossing. The difficulty, however, lies in the fact that continued parthenogenetic multiplication which is possible in Hydatina had the same effect as continued inbreeding. Shull introduces the interesting speculation that this similarity is due to a gradual adjustment of nucleus to cytoplasm during the asexual propagation—this being assumed to bring about the same results as a gradual approach toward homozygosis. We are inclined to attribute both changes to environmental causes, believing that if a proper change in diet had been made vigor would have been maintained.

While we are not justified in concluding from these experiments that inbreeding accompanied by rigid selection will be beneficial to bisexual animals, they certainly show close mating is not invariably injurious. They indicate that the results of inbreeding depend more upon the genetic composition of the individuals subjected to inbreeding rather than upon any pernicious influence inherent in the process itself; and, as will be emphasized more strongly later, it is a wholly different matter whether inbreeding results injuriously through the inheritance received, or whether consanguinity itself is responsible. Yet such a status for the problem is unsatisfactory. The experiments on animals bring to light no facts which may not be interpreted as the result of Mendelian factor recombination; but if one were to base his judgment on them alone, he could not truthfully make the didactic statement that *inbreeding per se* is *not injurious*. There would ever be the uncertainty with which the additional variable bisexuality always encumbers a genetic experiment. Fortunately, we may turn to the numerous experiments on hermaphroditic plants for the deciding vote.

Many wild species and cultivated varieties of plants are almost invariably self-fertilized, and apparently lack nothing in vigor, productiveness or ability to survive. Among wild plants many species of the family Leguminosæ, among cultivated plants—wheat, rice, barley, oats, tobacco, beans, tomatoes—are types characterized by very nearly continuous self-fertilization, and these plants are in no immediate danger of extinction.

On the other hand, the majority of the higher plants is provided with devices which promote natural cross-pollination, and show definite injurious effects when inbred artificially. Even species which are characteristically self-fertilized are crossed occasionally. This, together with the fact that nearly all plants and animals are benefited by crossing, led Knight as early as the close of the eighteenth century to believe self-fertilization is not a natural process and always produces more or less injurious results. His views were summed up in the statement, " nature intended that a sexual intercourse should take place between neighboring plants of the same species." Darwin, fifty years later, basing his conclusions upon observation of animals and direct experimentation with plants, was even more radical, and concluded that "nature abhors perpetual self-fertilization."

Darwin compared self-fertilized plants with intercrossed plants in many different species. In the majority of cases the self-fertilized plants were clearly inferior to the crossed plants. These facts led to the belief that the evil effects of inbreeding kept on accumulating until eventually a plant or animal continuously reproducing in that manner was doomed to extinction. His own experimental results came far short of proving such an assumption, however. The two plants with which inbreeding was practiced the longest—*Ipomea* and *Mimulus*—showed very little further loss of vigor after the first generation. What the experiments did show, most clearly, was segregation of the inbred stock into types differing in their ability to grow as well as in minor, visible, hereditary characters. In both species plants appeared which were superior to other plants derived from the same source, some being equal or even superior in vigor to the original cross-pollinated stock. The inbred plants differed from the original material most noticeably in the uniformity of visible characters. Darwin's gardener stated it was not necessary to label the plants, as the different lines were so distinct from each other and so uniform among themselves they could easily be recognized.

After several generations of inbreeding, Darwin found it made no difference in the resulting vigor whether the plants in an inbred lot were selfed or were crossed among themselves. This he correctly ascribed to the fact that the members of such an inbred strain had become germinally alike. With less justice he attributed this approach to similarity in inherited qualities to the fact that the plants were grown for several generations under the same conditions, but it is easy to see why he held so tenaciously to this view if one remembers the faith he had in the effect of environment on organisms. Such a view he deemed supported by the fact that crosses of selfed lines with the intercrossed lines (also inbred, but to a less degree) did not give as great increases in growth as crosses of selfed lines with fresh stock from other localities. His crosses between inbred lines did give noticeable increases in growth, however, in many cases equaling the original variety. This is well illustrated by *Dianthus,* in which the selfed line was crossed both with the intercrossed line and with a fresh stock. The ratio of each crossed population to the selfed population in height, number of seed capsules, and weight of seed produced is as follows:

	Selfed x Intercrossed	Selfed x Fresh Stock
Height, compared to selfed plants	100:95	100:81
No. capsules compared to selfed plants	100:67	100:39
Weight of seed compared to selfed plants	100:73	100:33

With Darwin we still attribute the greater increase of vigor in crosses of distinct stocks to a greater germinal diversity, but we differ from him as to the way in which that diversity is brought about. Be that as it may, great credit is due Darwin for being the first to see it was not the mere act of crossing which induced vigor but the union of different germinal complexes. This he states clearly in the following sentences ("Cross and Self-Fertilization in the Vegetable Kingdom," p. 269): "A cross between plants that have been self-fertilized during several successive generations and kept all the time under nearly

INBREEDING EXPERIMENTS

uniform conditions, does not benefit the offspring in the least or only in a very slight degree. Mimulus and the descendants of Ipomea, named Hero, offer instances of this rule. Again, plants self-fertilized during several generations profit only to a small extent by a cross with intercrossed plants of the same stock (as in the case of Dianthus), in comparison with the effects of a cross by a fresh stock. Plants of the same stock intercrossed during several generations (as with Petunia) were inferior in a marked manner in fertility to those derived from the corresponding self-fertilized plants crossed by a fresh stock. Lastly, certain plants which are regularly intercrossed by insects in a state of nature, and which were artificially crossed in each succeeding generation in the course of my experiments, so that they can never or most rarely have suffered any evil from self-fertilization (as with Eschscholtzia and Ipomea), nevertheless profited greatly by a cross with a fresh stock. These several cases taken together show us in the clearest manner that it is not the mere crossing of any two individuals which is beneficial to the offspring. The benefit thus derived depends on the plants which are united differing in some manner, and there can hardly be a doubt that it is in the constitution or nature of the sexual elements. Anyhow, it is certain that the differences are not of an external nature, for two plants which resemble each other as closely as the individuals of the same species ever do, profit in the plainest manner when intercrossed, if their progenitors have been exposed during several generations to different conditions."

Unfortunately, in Darwin's time the key to the solution of the problem of inbreeding was lacking. Mendel's work was yet unrecognized; the principles of inheritance of separate characters, of segregation, of chance recombination, Darwin was not permitted to know. Had he realized the way in which recessive characters can be concealed for many generations without making their appearance until homozygosity was brought about by inbreeding, doubtless his views on the subject would have been materially changed.

As we have just indicated, and as we shall have occasion to emphasize again, the greatest advance in our knowledge of the significance of inbreeding has come through linking its effects with Mendelian phenomena. The first experiments on the subject made in the light of this discovery were those of G. H. Shull and of East, undertaken independently in 1905 with maize, an ideal cross-fertilized species, as the subject.

Shull's investigations were not begun with the object of studying the effects of self-fertilization, but the studies having involved parallel cultures of cross-pollinated and self-pollinated lines, it was impossible not to have noticed the smaller stalks and ears and the greater susceptibility to attacks of the corn-smut (*Ustilago maydis*) shown by the latter. Interest thus aroused, data were collected bearing on the subject of inbreeding, and in 1908 his first conclusions on the subject were published.

His observation that the progeny of every self-fertilized maize plant is inferior in size, vigor and productiveness to the progeny of a normal cross-bred plant derived from the same source, corroborated preceding investigations made by Morrow and Gardner[152, 153] and

Shamel [192]; but the conclusion which he drew was new. The universality of this decrease in vigor was to Shull a proof that the injurious effect of inbreeding could not be due to an accumulation of deficiencies possessed by the parents since superior and inferior parents yielded similar results. Further, Shull noted that this decrease in size and vigor accompanying self-fertilization, instead of proceeding at a steady or even at an increasing rate as might be expected from this older view, actually became less and less in succeeding generations—presumably indicating an approach to stability. The neatness with which these observations fit a Mendelian interpretation of inbreeding did not escape notice. It was pointed out how one might consider a corn field to be a collection of complex hybrids whose elementary components may be separated by self-fertilization through the operation of the fundamental Mendelian laws of segregation and recombination.

With this working hypothesis the investigations were continued for several years, papers on the subject appearing in 1909, 1910 and 1911. Evidence of the hybrid nature of ordinary commercial maize plants and their dependence upon hybridity for their vigor was found in the decided differences in definite, hereditary, morphological characters exhibited by self-fertilized families having a common origin, but a further proof of the validity of the hypothesis came in testing the conclusions to which the view leads. Obviously crosses between plants of a single family, which by long-continued self-fertilization has become homozygous in nearly all its characters, should show little increase in vigor over self-fertilization; but crosses between distinct self-fertilized lines should often result in high-yielding F_1 generations possessing great vigor and showing a high degree of uniformity. Again, crosses between different near-homozygous strains, though uniform and vigorous in the F_1 generations, should become much more variable and much less vigorous in the F_2 generation. These general propositions Shull tested in a limited way in 1910 after his families had been self-fertilized for five generations. The variability of two such strains and the crosses between them for a definite and easily determined character—number of rows per ear—is shown in the following table:

Strain	Mean	Coefficient of variation
A	$8.30 \pm .06$	8.50 per cent. $\pm$ .47 per cent.
B	$14.10 \pm .15$	9.66 per cent. $\pm$ .74 per cent.
A$\times$B (F_1)	$12.71 \pm .15$	10.00 per cent. $\pm$ .87 per cent.
B$\times$A (F_1)	$11.77 \pm .07$	8.13 per cent. $\pm$ .42 per cent.
A$\times$B (F_2)	$11.84 \pm .11$	14.64 per cent. $\pm$ .67 per cent.
B$\times$A (F_2)	$13.79 \pm .11$	10.62 per cent. $\pm$ .56 per cent.

Clearly the F_1 generation, made with either type as the mother, is as uniform as the parent strains, but the F_2 generations are both more variable.

To test the other corollaries, nine different self-fertilized families of the fifth generation were compared with families obtained by crossing two plants belonging to each family; seven families were raised as first-generation hybrids between these different selfed strains; ten crosses between F_1 individuals were compared with ten self-fertilizations in the same families; and ten families were grown in which self-fertilization had been precluded for five years. The average height in decimeters, number of rows per ear, and yield in bushels per acre of these fifty-five families are given in the following table:

	Selfed	Selfed × Sibs	F_1	F_2	F_1 Selfed	F_1 Sib Crosses	Cross-breds
Average height..	19.28	20.00	25.00	23.42	23.55	23.30	22.95
Average No. rows	12.28	13.26	14.41	13.67	13.62	13.73	15.13
Average yield....	29.04	30.17	68.07	44.62	41.77	47.46	61.52

The sister-brother (sib) crosses give a slightly greater height, number of rows per ear and yield per acre than the corresponding self-fertilized families, an indication, as Shull states, of some heterozygosis still remaining in the selfed families; in other particulars Mendelian expectation is *wholly* confirmed.

The experiments of Shull on the effect of inbreeding in maize were continued only from 1905 to 1911. We may be pardoned, therefore, if we describe the experiments begun in 1905 by East at the Connecticut Agricultural Experiment Station in somewhat greater detail, for they are still being carried on by Jones. In fact, in point of numbers and scope they are the most extensive experiments on the problem of inbreeding. The general method of procedure has been merely to self-pollinate individual plants from different varieties of all the principal types of maize. The seed from such self-fertilized plants has been grown and some plants again self-fertilized. Thus a selfed plant has been the parent of each population. Over thirty different varieties, with several lines in each variety, have been inbred in this way. The oldest strains have now been self-fertilized for twelve consecutive generations.

In every case there has been a reduction in size of plant and yield of grain. Besides this result, to which there has been no exception, the several inbred lines originating from the same variety have become more or less strikingly differentiated in morphological characters. Some of the differences which characterize the several inbred strains in various combinations are as follows:

>Colored and colorless pericarps, cobs, silks and glumes.
>Profusely branched tassels and scantily branched or unbranched tassels.
>Long ears and short ears.
>Round cobs and flat cobs.
>Narrow silks and broad silks.
>Ears with various numbers of rows.
>Ears with straight rows and ears with irregular rows.
>Ears with large seeds and ears with small seeds.
>Ears high on the stalk and ears low on the stalk.
>Stalks with many tillers and stalks with few tillers.
>Leaves with straight margin and leaves with wavy margin.

Many other character differences governed by definite inherited factors have been observed, but these may serve as illustrations.

Along with these normal differences a number of characters have appeared which might well be called monstrosities, using the term not because of any abnormality in the method of their inheritance, but because they are not fitted to struggle for place either in agriculture or in nature. A common occurrence is the isolation of dwarf plants which are rarely capable of producing seed from their own pollen. Plants manifesting various degrees of chlorophyll deficiency are also frequently found. This may show in the form of an entire lack of chlorophyll, as seen in pure albino plants which live only until the supply of food in the seed is exhausted; or, it may appear as a yellowish green, the plants struggling through to seed production—though with some difficulty. Some plants are obtained with ear malformations and thus produce but a minimum amount of seed. Other plants lack brace roots and are unable to stand upright. Still others show various grades of pollen and ovule abortion, and susceptibility to disease.

The variability of the inbred lines in respect to the above characters decreased as inbreeding was continued. After four generations they were practically constant for the grosser characters. From the eighth generation on they have been remarkably uniform in all characters.

Inbreeding the naturally cross-pollinated maize plant, then, has these results:

1. There is a reduction in size of plant and in productiveness which continues only to a certain point and is in no sense an actual degeneration.

2. There is an isolation of subvarieties differing in morphological characters accompanying the reduction in growth.

3. As these subvarieties become more constant in their characters the reduction in growth ceases to be noticeable.

4. Individuals are obtained with such characters that they cannot be reproduced or, if so, only with extreme difficulty.

A large amount of data has been obtained upon which to base these statements, but since most of them have been published it seems desirable to include only a few illustrations here. The strains which have been the longest inbred will serve to show something as to the effect which inbreeding has had upon yield of grain, height of plant and other maize characters.

The original experiment began with four individual

plants obtained from seed of a commercial variety grown in Illinois known as Leaming Dent. This variety was given the number 1, and four plants which were self-pollinated and selected for continuation of the inbreeding experiment were numbered 1-6, 1-7, 1-9, and 1-12. These four lines were perpetuated each year by self-pollination and will be referred to hereafter as the Leaming strains. In the second inbred generation two self-pollinated plants in the 1-7 line were saved for seed and from them two inbred lines were split off which consequently came originally from one line inbred two generations. These were numbered 1-7-1-1 and 1-7-1-2. Many other inbred strains coming from different material have been started from time to time and several of them are still being continued. There is no need to mention them specifically, except as they bring out special features.

TABLE III
The Effect of Inbreeding on the Yield and Height of Maize

Year grown	No. of generations selfed	Four inbred strains derived from a variety of Leaming dent corn							
		1-6-1-3-etc.		1-7-1-1-etc.		1-7-1-2-etc.		1-9-1-2-etc.	
		Yield bu. per acre	Height inches	Yield bu. per acre	Height inches	Yield bu. per acre	Height inches	Yield bu. per acre	Height inches
1916	0	74.7	117.3	74.7	117.3	74.7	117.3	74.7	117.3
1905	0	88.0	.....	88.0	.....	88.0	.....	88.0	.....
1906	1	59.1	.....	60.9	.....	60.9	.....	42.3	.....
1908	2	95.2	.....	[1907]59.3	.....	[1907]59.3	.....	51.7	.....
1909	3	57.9	.....	[1908]46.0	.....	[1908]59.7	.....	35.4	.....
1910	4	80.0	.....	63.2	.....	68.1	.....	47.7	.....
1911	5	27.7	86.7	25.4	81.1	41.3	90.5	26.0	76.5
1912	6	.....	.....	.....	.....	.....	.....	[1913]38.9	.....
1913	7	41.8	.....	39.4	.....	.....	.....	[1914]45.4	85.0
1914	8	78.8	96.0	47.2	83.5	58.5	88.0	[1915]21.6	.....
1915	9	25.5	.....	24.8	.....	.....	.....	[1916]30.6	78.7
1916	10	32.8	97.7	32.7	84.9	19.2	86.9	[1917]31.8	82.4
1917	11	46.2	103.7	42.3	78.6	37.6	83.8	.....	.....

In Table III the yield of grain and height of plant of the four inbred Leaming strains are given in the successive generations of self-fertilization. In 1916 seed of the original variety, which had been grown in the meantime in the locality in Illinois from whence it was originally secured, was obtained and grown for comparison with the inbred strains. This variety in Illinois in 1905 yielded at the rate of 88 bushels of shelled grain per acre and in Connecticut in 1916 at the rate of 75 bushels. There is no reason for supposing that the variety had changed to any great extent in the intervening years. Coming from Illinois, it was placed at a disadvantage as compared to the inbred strains, because it was not adapted to the local conditions, while the inbred strains, grown for several years, had been selected more or less unconsciously to meet the prevailing conditions. Even with this in favor of the inbred strains they yielded only from one-third to one-half as much as the original variety grown under the same conditions.

With regard to the rate of reduction in yield or the constancy of the varieties during the later generations, it is difficult to draw conclusions from these figures, owing to the fluctuation in yield from year to year due to seasonal conditions and to the difficulty of accurate testing in field plot work, a fact recognized by all who have made such tests. No yields for any of the strains were taken in 1912. The yields for 1909 and 1915 were too low on account of poor seasons. The yields in 1914 were too high for the opposite reason. In 1915 the yields were unreliable because only a few plants were available for calculation, most of the plants having been used for hand pollinations.

In 1916 and 1917 the inbred strains were grown in somewhat larger plots and the yields are fairly reliable.

With these points in mind, an examination of the table shows that from the beginning of the experiment to the ninth generation there has been a tremendous drop in productiveness, so that in that generation the strains were approximately only one-third as productive as the variety before inbreeding. From the ninth to the eleventh generation there has been no reduction in yield and practically no change in visible characters. Height of plant, as far as the available figures show, followed the same course. The reduction which has taken place occurred in the first eight generations; after that there has been no appreciable change.

All along the several Leaming strains have shown considerable differences in productiveness and in height. Strain No. 1-6 has given the largest yields and the tallest plants. It gave nearly 50 per cent. larger yields than the poorest yielding strain in the eleventh year, and was about 30 per cent. higher than the shortest strain.

One of the strains, No. 1-12, was lost in the sixth generation. Previous to this time it had been the poorest of the five. It was partially sterile, never produced seed at the tip of the ear and was perpetuated only with care. Since the difficulty of carrying along any inbred strain is great, owing to failure to pollinate at the correct time, to attacks of fungus on the ears enclosed in paper bags, and to poor germination in the cold, wet weather common in New England at corn-planting time, the loss of this strain might be easily accounted for without assuming continuous deterioration. The strain probably could have been retained if sufficient effort had been put forth; but in view of the further reduction in other strains, it would have been extremely difficult. Since plants are frequently produced which cannot be perpetuated, however, it is to be expected that some strains will also be found which cannot survive. This is good evidence that strains, differing markedly in their ability to grow, are isolated by inbreeding.

Plants of the surviving strains, while smaller in size and lower in productiveness, are perfectly healthy and functionally normal in every way except that in many of them there is an extreme reduction in the amount of pollen produced. These infertile types are dependent on other plants for pollen in order to make the yields they show in open field culture; when grown by themselves the yield is less due to an inadequate supply of pollen. On the other hand, this extreme reduction in pollen production is not shown by all the strains, some inbred strains producing pollen abundantly.

From the data given in Table III there is considerable evidence that these plants have reached about the limit of their reduction in size and productiveness and that whatever changes have taken place in the last three years have been slight. Further inbreeding is necessary for one to be positive on this point. But as the crosses within these inbred strains have given no significant increases over the selfed lines, and as there has been no visible change in morphological characters, in the past three years at least, it seems apparent that the reduction in vegetative vigor and productiveness is very nearly, if not quite, at an end.

Reduction is shown by inbred maize plants in other characters. Length of ear, as well as height of plant and yield of grain, is smaller. There is also a slight reduction in number of nodes and in rows of grain, but in contrast to the other three characters the change is almost negligible. The last two are only slightly affected by environmental factors as compared with the others. A plant may be reduced to one-half its normal height by being grown in a poor situation, but the number of nodes will be nearly the same in the two cases. Hence, we see that inbreeding affects plants much in the same way as poor environmental conditions.

In all of the characters mentioned there is a reduction in variability and change in mean differing in the several lines. This is illustrated in Table IV, in which are given the data for number of rows of grain on the ear of four different plots of the original non-inbred variety and four strains derived from this variety after ten generations of self-fertilization. The marked reduction in variability is apparent both in the restricted range of the distribution of the inbred lines compared to the variety, and in the coefficients of variability.

This reduction in variability applies only to each inbred line separately. If all the different lines were combined together into one population the variation would be greater than that shown by the original material. This is readily apparent from the table; it also follows from the fact that many characters are produced by inbreeding which are seldom seen in the regularly cross-pollinated stock. Inbreeding reduces variability within separate lines, but increases variability in the descendants as a whole.

From the curves on inbreeding given in the preceding chapter (Fig. 24), it was seen that the production of com-

INBREEDING EXPERIMENTS

TABLE IV

Frequency Distribution of the Number of Rows of Grain on the Ear of a Non-Inbred Variety of Maize and Inbred Strains Derived From It

Pedigree number	Number of rows on the ear								N.	Ave.	C. V.
	12	14	16	18	20	22	24	26			
1 (Original)	..	2	14	18	14	5	2	2	57	18.7±.23	14.12±.91
1 non-inbred	..	1	15	21	8	3	3	..	51	18.2±.21	12.42±.84
1 Learning	..	6	7	19	17	7	2	2	60	18.4±.23	14.67±.93
1 variety	..	5	15	17	13	6	0	2	59	18.2±.24	15.16±.96
Total	3	14	51	75	52	21	7	4	227	18.4±.12	14.22±.46
1-6-1-3-4-4-4-2-5-3-1	6	37	13	..	..	..	..	..	56	14.3±.10	7.96±.50
1-6-1-3-4-4-4-2-5-5-1	5	34	7	..	..	..	..	..	46	14.4±.10	7.23±.51
1-6-1-3-4-4-4-1-2	1	12	42	4	..	..	..	..	58	15.7±.09	6.48±.40
1-6-1-3-4-4-4-2-6-1	..	7	24	22	6	..	..	..	59	16.9±.14	9.83±.61
1-7-1-1-4-7-5-2-6-1	..	..	..	5	15	30	12	2	62	21.8±.16	8.43±.51
1-7-1-1-4-7-5-2-1-1	..	..	..	6	14	19	14	3	56	22.0±.20	8.10±.64
1-7-1-1-4-7-5-4-5-2	..	..	1	10	13	11	1	..	36	20.1±.20	8.94±.71
1-7-1-1-4-7-5-4-7-2	..	..	..	1	16	23	11	1	52	21.8±.15	7.48±.49
1-7-1-2-2-9-2-1-1-3	..	12	20	10	..	..	..	..	42	15.9±.15	9.06±.66
1-7-1-2-2-9-2-1-1-4-1	..	12	22	8	1	..	..	..	43	15.9±.15	9.40±.68
1-9-1-2-4-6-7-5-3-3	..	19	30	3	2	..	..	..	55	15.5±.13	9.21±.59
1-9-1-2-4-6-7-5-6-4	..	20	40	..	..	..	..	..	59	15.4±.08	6.08±.38

pletely homozygous types by self-fertilization is greatest in the generations from the third to the sixth if a large number of factor differences are involved at the start. The experimental results obtained from these inbred strains of maize fit this theory well. It is not until after about three generations of self-fertilization that extreme types begin to appear. While there has been a reduction in size and productiveness before this, it is at this time, or during the next two or three generations, that the greatest diversity of types occurs. It is here that most of the monstrosities and plants which are unable to reproduce themselves appear.

From Table IV we see that equally striking changes in the mean row number also take place. The averages have been shifted both up and down from the original conditions. The greatest segregation has taken place between the first and the eighth generations. In the eighth generation the lines were again split up, but show no marked change after this point. Differences in the ears of these inbred strains of corn are shown in Fig. 29.

The rate of reduction in variability and rate of change of mean are shown by the data for row number of two of the inbred strains for successive years in Table V and Fig. 30. These two lines are descended from the same plant in the second year of self-fertilization. The figures previous to the third year are not available, and in that year only for one of the strains, but since then a marked change in average row number, and a reduction in variability have taken place without conscious selection one way or the other. Though the number of plants grown in the generations from the 7th to the 10th are too few to be a basis for accurate conclusions, the sharp increase in

TABLE V
REDUCTION IN VARIABILITY AND SEGREGATION OF EAR ROW NUMBER IN INBRED STRAINS OF MAIZE

Year grown	Generations selfed.	Pedigree number.	12	14	16	18	20	22	24	26	28	N.	Ave.	C. V.
1908	3	1-7-1-1	3	3	4	6	27	7	4	1	2	57	19.7 ± .30	17.00 ± 1.10
1908	3	1-7-1-2												
1910	4	1-7-1-1-1			2	5	11	17	19	6	1	61	20.2 ± .22	12.67 ± .78
1910	4	1-7-1-2-2				2	8	14	9	5	1	39	20.5 ± .25	11.20 ± .86
1911	5	1-7-1-1-4				3	4	10	16	4	1	65	20.9 ± .25	10.92 ± .85
1911	5	1-7-1-2-9				14	29	21	1			65	18.3 ± .13	8.34 ± .49
1912	6	1-7-1-1-4-7												
1912	6	1-7-1-2-9-2												
1913	7	1-7-1-1-4-7-5			1	3	5	1				10	19.2 ± .34	8.33 ± 1.26
1913	7	1-7-1-2-9-2-1		4	12	2						18	15.8 ± .18	7.16 ± .80
1914	8	1-7-1-1-4-7-5-4			1	9	16	6				32	21.7 ± .18	4.72 ± .40
1914	8	1-7-1-2-9-2-1-1			17	20	6					43	17.5 ± .14	7.83 ± .57
1915	9	1-7-1-1-4-7-5-4-5			1	14	4	2				21	20.7 ± .21	6.90 ± .72
1915	9	1-7-1-2-9-2-1-1-4		6	16	7						29	16.1 ± .17	8.33 ± .74
1916	10	1-7-1-1-4-7-5-4-5-2		1	10	13	11	1				36	20.1 ± .20	8.94 ± .71
1916	10	1-7-1-2-9-2-1-1-4-3		12	22	8	1					43	15.9 ± .15	9.40 ± .68
1917	11	1-7-1-1-4-7-5-4-5-2-1		3	23	45	20	43	22	7		95	20.2 ± .13	9.15 ± .45
1917	11	1-7-1-2-2-9-2-1-1-4-3-1	3	23	45	22	2					95	15.9 ± .11	10.39 ± .51

average row number and decrease in variability in the eighth generation are probably due to the favorable growing conditions of that year—witness the high yields for the inbred strains in that year as given in Table III. The apparent rise in variability after the eighth generation is in part due to the fact that the ears had become some-

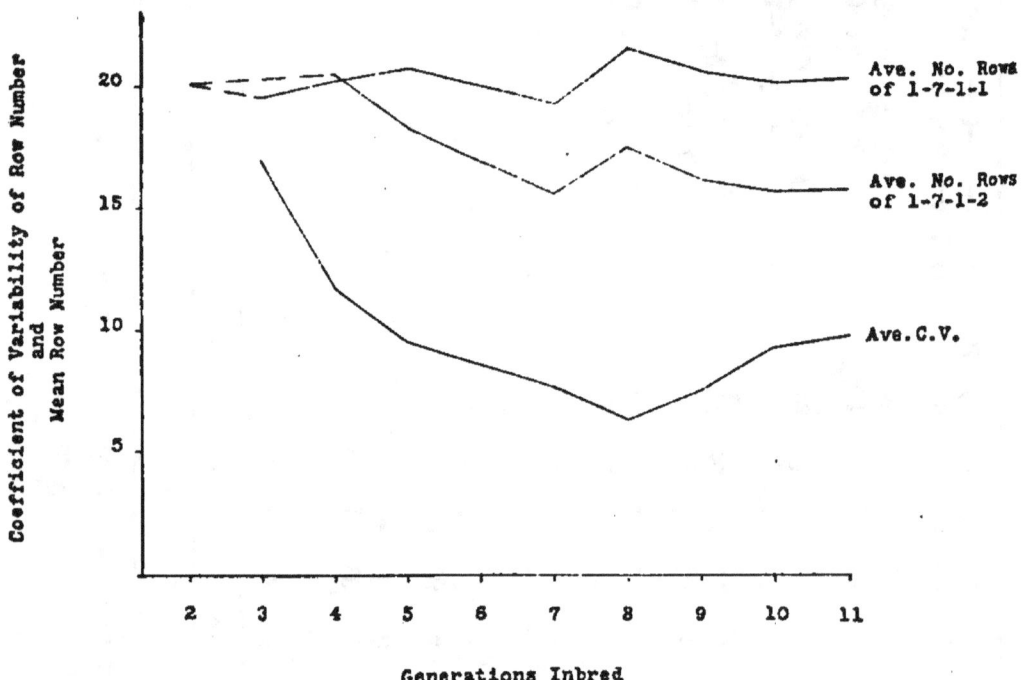

Fig. 30.—Graphs showing reduction of variability and segregation of ear row number in selfed strains of maize.

what more irregular in row number, so that accurate determination of number of rows has been more difficult in the later generations. However, this rise is more apparent than real as the values for the coefficients of variability in the intermediate generations are probably lower than they would have been if an adequate number of plants had been grown.

The effect of inbreeding upon variability is even more

apparent in details of plant and ear structure which are difficult of statistical expression. The beautiful uniformity of these plants in all characteristics at the present time is one of their most striking features. This can be seen fairly well for the ear characters in accompanying illustrations (Fig. 29). In the minutiæ of the tassels, leaves and stalks they show the same striking uniformity. These minor details which characterize each of these groups of plants are difficult to describe adequately, but are perhaps the most noticeable feature about them. The tassels or the ears of all these four Leaming strains, if mixed together, could be separated without the slightest difficulty.

Some characters appear so rarely in plants they have been generally considered to be due to what might be called physiological accidents rather than to inheritance. An illustration of this kind is furnished in maize by the occurrence of doubled or connate seeds. Instead of one embryo enclosed in a pericarp, separate embryos and endosperms are present, with the seeds arranged back to back and the embryos facing in opposite directions. A few seeds of this kind have been described from time to time, but never more than one or two on an occasional ear. From twelve inbred strains of a variety of maize other than the ones previously described, two lines have been obtained which produce these peculiar seeds as a common feature. One of the strains shows from one to six or more on practically every ear. The second strain shows them more rarely and the other ten strains derived from the same variety have never been observed to bear them. Here, then, is a character which does not appear except at rare intervals when the plants are crossed and in full vigor. When the plants are brought to homozygosity and the vigor of the plants is reduced, the doubled seeds appear in abundance in some lines, but not in all. A character, then, may be governed in its expression by other characters and modified by the vigor of the plant, but in the final analysis it is dependent upon definitely inherited factors.

In the same way such indefinite and complex characters as susceptibility and resistance to disease are shown to be capable of segregation. In 1917 one of the inbred Leaming strains had not a single plant affected by the smut fungus, although 1000 plants were grown in different places. Other strains derived from the same variety and grown side by side with the susceptible race showed from 5 to 10 per cent. of plants infected. Susceptibility of maize to smut thus seems to be dependent upon inherited factors. As the result of inbreeding, these factors may be segregated into some lines and not into others.

Although there has been a striking reduction in size of plant, general vegetative vigor and productiveness, and in comparison with non-inbred varieties the inbred plants are more difficult to grow, emphasis must be put upon the fact that they are normal and healthy. No actual degeneration has occurred. The monstrosities which are common in every field of maize, such as the occurrence of seeds in the tassels, anthers in the ears, dwarf plants, completely sterile plants, and other similar anomalies, now no longer appear in these inbred strains. These facts, taken together, should be sufficient to demonstrate beyond doubt that by far the greatest amount of the general variability found among ordinary cross-fertilized plants is due to the segregation and recombination of definite and constant hereditary factors. Some of the characters which appear after long-continued inbreeding are seldom seen in continually cross-pollinated plants, and never are so many seen in combination. This is because they are recessive in nature and complex in mode of inheritance. The most significant feature about the characters which make their appearance in inbred plants is that none of them can be attributed directly to a loss of a physiological stimulation, although undoubtedly many of them may be modified by the vigor of the plants upon which they are borne. There is no one specific feature common to all inbred strains, but simply a general loss of vigor, a general reduction in size and productiveness accompanied by specific characters more or less unfavorable to the plant's best development. But these unfavorable characters are never all found in one inbred strain, nor is any one of them found in all inbred strains.

Although no systematic selection has been practiced throughout these inbreeding experiments, a great deal of selection upon many characters has been unavoidable as is the case in any inbreeding experiment. In maize, the difficulties of hand pollination result in the selection of plants whose staminate and pistillate parts are matured synchronously. Any great difference in this respect, particularly towards protandry, renders self-fertilization difficult or impossible as the pollen is viable but a short time. Of course, all plants which are weak, sterile, diseased or in any way abnormal, tend to become eliminated wherever these causes reduce the chance of obtaining seed. This unconscious selection becomes more rigid in the later generations of inbreeding as reduction in vigor and productiveness becomes more pronounced. Again, the small amount of seed produced by hand pollination under the most favorable circumstances, necessitates the using of the best ears obtained for planting in order to have enough plants upon which to make any fair observations.

These factors tend to prevent the attainment of complete homozygosity. Nevertheless, all the evidence at hand indicates that the four strains of Leaming corn which have been continuously self-fertilized for twelve generations are now very nearly, if not completely, homozygous in all inherited characters. As stated before, this evidence comprises cessation of reduction in size and productiveness, of reduction in variability, and of change of average row number and other characters. But there are still other ways of testing the proposition. On the theory that increase in growth results from crossing when the individuals united differ in respect to some inherited qualities, if no increase results, then the parents have no differences. These strains have been tested in this way by crossing different plants within a strain and comparing the crossed plants with selfed plants. While some increases in growth resulted from such crossing they were balanced by decreases in other cases, so that the inconsistencies are most likely due to difficulty in securing an accurate test. At the same time one should not shut his eyes to the possibility that some of the strains have reached complete homozygosity, while others, as yet, have not; although no sure evidence of such a state of affairs has been obtained.

Most of the direct experimentation to determine the effects of inbreeding has been with cultivated plants and domestic animals. The question will undoubtedly be asked, therefore, as to whether the results would have been the same had wild species been investigated. It would be

INBREEDING EXPERIMENTS

futile to maintain that there is every reason to suppose wild species should behave exactly as their domestic cousins. Wild types, in general, might not present such an appearance of injury under inbreeding as is often shown by cultivated species. This would not be due to differences in their method of inheritance, however, but because wild species are usually exposed to a more rigorous struggle for existence and the individuals are, therefore, less likely to differ by a large number of hereditary factors. For such reason one should expect experiments on different wild species to give rather varied results, and in the comparatively small number which have been made this is the case. Castle's experiments on the fruit fly gave no markedly unfavorable results. Collins states that self-fertilizing teosinte, a semi-wild relative of maize, causes no loss of vigor. Yet Darwin compared self-fertilized and intercrossed plants of several species which are largely cross-fertilized in the wild with great disadvantage to the former.

This discussion of the effects of artificial inbreeding in certain plants and animals has been given in some detail in order to bring out the many important considerations involved. There has even been repetition in order to emphasize the most important points. Details are merely by way of parenthesis, however. Let us now get out of the parenthesis and into the main argument.

From the preceding observation it can be said that inbreeding has but one demonstrable effect on organisms subjected to its action—the isolation of homozygous types. The diversity of the resulting types depends directly upon number of heterozygous hereditary factors present in the individuals with which the process is be-

gun; it is likely, therefore, to vary directly with the amount of cross-breeding experienced by their immediate ancestors. The rapidity of the isolation of homozygous types is a function of the intensity of the inbreeding.

Take the case of maize as an example. Maize is one of the most variable of cultivated plants, and is usually cross-pollinated under natural conditions. In other words, the individuals making up any commercial variety of maize are each and every one heterozygous for a large number of hereditary factors—a heterozygosis that is kept up by continual crossing and recrossing. When such a variety is inbred there is automatic isolation of homozygous combinations, following simple mathematical laws as we have already seen. If self-fertilization is practiced, stabilization through an approximately complete homozygosis occurs after a relatively small number of generations; if a less intense system of inbreeding is followed, the result is the same, but it is obtained more slowly. During this process, before stabilization is reached, there is reduction in size, vigor and productiveness following somewhat roughly the reduction in per cent. of heterozygousness. We can think of this reduction in vigor as a change correlated with approaching homozygosis if we wish, although as we shall see there is reason to believe it to be a result of linked inheritance. What does occur is a reduction in vigor of the population as a whole in each generation associated with the isolation of individuals *more homozygous than their parents.* Any particular individual may be vigorous or weak, fertile or sterile, normal or monstrous, good, bad or indifferent, depending wholly upon the combination of characters received. Many of the characters which become homozygous will be recessives or combinations of recessives which seldom are seen under ordinary circumstances, because they are hidden by their dominant allelomorphs. These recessives are the "corrupt fruit" which give the bad name to inbreeding, for they are often—very often—undesirable characteristics.

The homozygous inbred strains after stability has been reached are quite comparable to naturally self-fertilizing species provided they have passed as rigorous selection as the latter have had to undergo by reason of natural competition. And Darwin, as well as others, found that artificial self-pollination causes no reduction in such genera as Nicotiana, Pisum and Phaseolus where self-fertilization is the general rule.

Are then the immediate results of inbreeding sometimes injurious? In naturally cross-fertilized organisms they most emphatically are—nay, more, even disastrous—when we recall the reduction to over half or one-third in production in grain and a corresponding decrease in size of plant and rate of growth in maize. But maize is probably an extreme case. With other organisms the results are not so bad, and in some cases, especially when selection has been made, no evil effects are apparent. In fact, there may be an actual improvement. But the truth is, we did not set out to answer that question. It had already received a correct answer. *What we undertook to inquire was whether inbreeding is injurious merely by reason of the consanguinity.* We answer, *No!* The only injury proceeding from inbreeding comes from the inheritance received. The constitution of the individuals resulting from a process of inbreeding depends upon the chance allotment of characters preëxisting in the stock before inbreeding was commenced. If undesirable characters are shown after inbreeding, it is only because they already existed in the stock and were able to persist for generations under the protection of more favorable characters which dominated them and kept them from sight. The powerful hand of natural selection was thus stayed until inbreeding tore aside the mask and the unfavorable characters were shown up in all their weakness, to stand or fall on their own merits.

If evil is brought to light, inbreeding is no more to be blamed than the detective who unearths a crime. Instead of being condemned it should be commended. After continued inbreeding a cross-bred stock has been purified and rid of abnormalities, monstrosities, and serious weaknesses of all kinds. Only those characters can remain which either are favorable or at least are not definitely harmful to the organism. Those characters which have survived this "day of judgment" can now be estimated according to their true worth. As we shall see later vigor can be immediately regained by crossing. Not only is the full vigor of the original stock restored, but it may even be increased, due to the elimination of many unfavorable characters. If this increased vigor can be utilized in the first generation, or if it can be fixed so that it is not lost in succeeding generations, then inbreeding is not only not injurious but is highly beneficial. As an actual means of plant and animal improvement, therefore, it should be given its rightful valuation.

CHAPTER VII

HYBRID VIGOR OR HETEROSIS

Whether or not inbreeding in a race of plants or animals results injuriously depends primarily, as we have attempted to show, upon the hereditary constitution of the organism. The beneficial effect of crossing, heterosis, is a more widespread phenomenon. It may be expected when almost all somewhat nearly related forms are crossed together. Even plants or animals which show no harmful results of inbreeding are frequently improved thus in a remarkable way. Moreover, this stimulating effect is immediately apparent in the individuals resulting from the cross. It is then at its maximum.

It is natural, therefore, that the early writers on the subject should have noticed and emphasized the good to be derived from crossing rather than the bad which sometimes results from inbreeding. Almost without exception the great horticultural writers of the late eighteenth and early nineteenth centuries noted the occurrence of hybrid vigor, and many of them described it in great detail. Among them may be mentioned Kölreuter (1763), Knight (1799), Mauz (1825), Sageret (1826), Berthollet (1827), Wiegmann (1828), Herbert (1837), Lecoq (1845), Gärtner (1849). In fact, in Focke's compilation of this early work, "Die Pflanzen-Mischlinge" (1881), cases of heterosis worthy of special mention were found in fifty-nine families of the flowering plants as well as in the conifers and the ferns. Animal husbandmen were somewhat less inclined to acknowledge and discuss the matter, although they had an excellent example before them in the mule—an animal known and appreciated for over four thousand years. But the necessity of their following the custom of maintaining breeds true to certain fixed standards probably accounts for their conservatism in estimating the importance of the phenomenon.

Kölreuter,[125] the first botanist to study artificial plant hybrids, made many interspecific crosses in the genera Nicotiana, Dianthus, Verbascum, Mirabilis, Datura and others, which astonished their producer by their greater size, increased number of flowers and general vegetative vigor, as compared with the parental species entering into the cross. He gives many exact measurements of his hybrids and speaks with some awe of their *"statura portentosa"* and *"ambitus vastissimus ac altitudo valde conspicua."* Later, after some observations on certain structural adaptations for cross-pollination which he interpreted correctly, he made a passing remark which plainly showed he thought Nature had intended plants to be cross-fertilized and that benefit ensued therefrom.

Some forty years after, Thomas Andrew Knight,[122] a horticulturist who was a very keen observer, noticed similar instances of high vigor in his crosses: in the description of these experiments we note the following remarks concerning a cross between two varieties of peas:

By introducing the farina of the largest and most luxuriant kinds into the blossoms of the most diminutive and by reversing the process I found that the powers of the male and female in their effects on the offspring are exactly equal. The vigor of the growth, the size of the seeds produced, and the season of maturity, were the same, though the one was a very early, and the other a very late variety. I had, in this experiment, a striking instance of the stimulating effects of crossing the breeds; for the smallest variety, whose height rarely exceeded two feet, was increased to six feet, whilst the height of the large and luxuriant kind was very little diminished.

It is evident that in this particular case Knight was dealing with dwarf and standard peas, and dominance of the tall standard habit of growth is to be expected. This is not the correct interpretation of the majority of his observations on hybrid vigor, however; a sufficient number of really striking manifestations of the phenomenon were found to give adequate foundation for his anti-inbreeding principle, elaborated by Darwin fifty years later.

Probably the most extensive series of early experiments on hybridization were those of Gärtner.[74] This enthusiastic worker crossed, or attempted to cross, everything available to him. According to Lindley,[126] he made 10,000 pollinations between 700 species, and produced 250 different hybrids. Many of his attempted crosses either failed to produce seed, or if seed was produced, gave feeble plants; but a great number of the hybrids, where the crosses were made between plants not too distantly related, showed distinct evidence of hybrid vigor manifested in many different ways. Gärtner speaks especially of their general vegetative luxuriance, increase in root development, height, number of flowers, the facility of their vegetative propagation, their hardiness and early and prolonged blooming. He says:

> One of the most conspicuous and common characteristics of plant hybrids is the luxuriance of all their parts, a luxuriance that is shown in the rankness of their growth and a prodigal development of root shoots, branches, leaves, and blossoms that could not be induced in the parent stocks by the most careful cultivation. The hybrids usually reach the full development of their parts only when planted in the open, as

Kölreuter (125) has already remarked; when grown in pots and thus limited in food supply their tendency is toward fruit development and seed production.

Besides possessing general vegetative vigor, hybrids are often noticeable for the extraordinary length of their stems. In various hybrids of the genus Verbascum, for example *lychnitis-thapsus*, the stem shoots up 12 to 15 feet high, with a panicle 7 to 9 feet, the six highest side branches 2 to 3 feet, and the stem 1 1/4 inches in diameter at the base: in *Althaea cannabino-officinalis* the stem is 10 to 12 feet; in *Malva mauritano-sylvestris* 9 to 11 feet; in *Digitalis purpureo-ochroleuca* 8 to 10 feet, with panicles 4 to 5 feet; and in *Petunia nyctaginifloro-phœnicea* and *Lobelia cardinali-syphilitica* 3 to 4 feet each. Prof. Wiegmann also corroborates these observations.

The root system and the power of germination of hybrids are highly correlated with their great vegetative vigor. Many hybrids, therefore, which are not so luxuriant in growth as those just described, for example Dianthus, Lavatera, Lycium, Lychnis, Lobelia, Geum, and Pentstemon hybrids, put forth stalks easily and therefore are readily propagated by layers, stolons, or cuttings. The observations of Kölreuter (125), and of Sageret (191) agree with ours in this respect.

Luxuriation expresses itself at times as proliferation; for instance, in *Lychnis diurno-flos cuculi* the receptaculum is changed to a bud that puts forth branches and leaves. If, moreover, the vigor of the hybrids especially affects the stem and the branches, particularly their length, nevertheless the leaves take part in it by becoming larger. Hybrids in the genera Datura, Nicotiana, Tropæolum, Verbascum, and Pentstemon are examples.

Naudin,[159] the contemporary of Mendel, whose ideas very nearly resembled modern conception of heredity, likewise gives many excellent illustrations of hybrid vigor from interspecific crosses which he made in Papaver, Mirabilis, Primula, Datura, Nicotiana, Petunia, Digitalis, Linaria, Luffa, Coccinea and Cucumis. Out of 35 crosses within these genera 24 show positive evidence of heterosis. The cross of *Datura Stramonium* with *D. Tatula* was particularly notable in this respect. Both reciprocal hybrids were twice as tall as either parent.

HYBRID VIGOR OR HETEROSIS 145

Even Mendel's classic pea hybrids supply further instances of increase in size resulting from crossing. Concerning them, he says:

> The longer of the two parental stems is usually exceeded by the hybrid, a fact which is possibly only attributable to the greater luxuriance which appears in all parts of the plants when stems of very different lengths are crossed. Thus, for instance, in repeated experiments, stems of 1 foot and of 6 feet in length yielded without exception hybrids which varied in length between 6 feet and 7 1/2 feet.

Focke,[70] in the book already cited, gives the results of a series of experiments nearly as extensive as those of Gärtner and catalogues his own results along with those of his predecessors. The compilation is so careful, so painstaking, and so complete that one may turn to the final conclusions of the author without fear of error as far as the facts are concerned. He says: "Crosses between different races and different varieties are distinguished from individuals of the pure type, as a rule, by their vegetative vigor. Hybrids between markedly different species are frequently quite delicate, especially when young, so that the seedlings are difficult to raise. Hybrids between species or between races that are more nearly related are, as a rule, however, uncommonly tall and robust, as is shown by their size, rapidity of growth, earliness of flowering, abundance of blossoms, long duration of life, ease of asexual propagation, increased size of individual organs, and similar characters."

The attention of these earlier hybridizers was mainly directed towards interspecific crosses, but they also noted a great number of instances in which crosses between closely related forms, such as varieties or sub-varieties of cultivated plants, gave remarkable increments in growth. In fact, we have found no record of intervarietal crosses where delicate or weak progeny resulted. It would not be useful, however, to attempt to canvass the literature for all those cases in which crossing either did or did not result to the advantage of the offspring. A list of the crosses would alone fill a volume. It is only necessary to point out that the value to be derived from crossing thus made so evident gave great impetus to the study of floral structures as adaptations for cross-pollination. So zealously was this line of investigation pursued, that knowledge of the methods of pollination in the angiosperms soon exceeded that of any other phase of general botany. The interpretation placed upon many of these floral mechanisms was fantastic, to say the least, the enthusiastic claims of the workers rivalling those of zoölogists in mimicry and protective coloration. The net result was simply to show how widespread were means of cross-pollination. It might be said to have proved that cross-fertilization is an advantage; it did not prove it to be indispensable. There were too many naturally self-fertilized plants for any such conclusion.

Of all the work on the effects of crossing in pre-Mendelian times, that of Darwin is the most important. With it we get a new insight into the meaning of inbreeding and outbreeding. Darwin was the first to see it was not the mere act of crossing which was beneficial. He satisfied himself on this point by crossing different flowers on the same plant and different plants of similar strains. In neither case was there any positive evidence of an effect. But crosses between different varieties or species of plants gave unmistakable signs of invigoration. In 24 cases out of 37, cross-fertilization increased the height of plant; in 5 out of 7 experiments, the weight was increased. Moreover, the crossed plants frequently flowered earlier and in many other ways showed their advantage over the parent races.

Darwin extended his observations to the animal kingdom and his views on the whole subject are summed up concisely in the following paragraph from "Animals and Plants under Domestication": "The gain in constitutional vigor derived from an occasional cross between individuals of the same variety, but belonging to different families, or between distinct varieties, has not been so largely or so frequently discussed as have the evil effects of too close interbreeding. But the former point is the more important of the two, inasmuch as the evidence is more decisive. The evil results from close interbreeding are difficult to detect, for they accumulate slowly and differ much in degree with different species, whilst the good effects which almost invariably follow a cross are from the first manifest. It should, however, be clearly understood that the advantage of close interbreeding, as far as the retention of character is concerned, is indisputable and often outweighs the evil of a slight loss of constitutional vigor."

From this statement Darwin evidently considered the ill effects of inbreeding and the good effects of crossing to be two different things. He was right in stressing the benefit from crossing rather than the injury from close mating, but wrong in thinking the evil effects accumulated as inbreeding was continued. Such a belief is not substantiated by more recent experiments, as has been shown in the last chapter. It is true, however, that the effect of inbreeding may not be as noticeable in the first generation as the invigoration immediately apparent after crossing.

The effects of outbreeding, unlike those of inbreeding, are shown both by plants which are naturally self-fertilized and by those which are cross-fertilized. Many of the illustrations already given are from plants almost invariably self-fertilized. Crossing within a pure line of such a species shows no heterosis; but if the parents united in the cross differ more or less in minor external features an increase in growth is usually to be expected. This has been shown to be true for peas, tomatoes, tobacco and many other normally self-fertilized forms among cultivated plants, as well as for several wild species.

An extensive series of crosses between different Nicotiana species has been reported by East and Hayes.[59] The majority of these crosses were taller than the average of the two parents and many were taller and more vigorous than either parent. Some of the crossed plants were completely sterile. In certain cases these were weak, non-vigorous plants, but there were others in which inability to produce seed was accompanied by increased vigor. Thus, while occasionally the increased development of sterile hybrids may be due in part to their having expended no energy in seed production, the fact that many vigorous hybrids manifest greater ability to produce seed shows this is a relatively unimportant factor and entirely inadequate to account for the great vigor obtained where there is full fertility.

A fair example of the way in which height is gained by crossing is given by East and Hayes,[59] a cross of *Nicotiana rustica brazilia* Comes and *N. rustica scabra* Comes.

The frequency distributions of height of plant of the two parents and the reciprocal hybrids are given in Table VI.

TABLE VI

Height of Crosses Between Nicotiana Rustica Scabra (352) and N. Rustica Brazilia (349)

Variety or cross	Class centers in inches																	
	24	27	30	33	36	39	42	45	48	51	54	57	60	63	66	69	72	75	78
349	4	10	22	14	7														
352							2	1	5	11	16	17	6						
352×349, F$_1$										1	3	0	5	5	5	6	1	1	
349×352, F$_1$												3	5	2	4	6	5	1	2

In both of the first hybrid generations the average height is above the major extreme of either parent. Similar increases in height were obtained when a commercial variety of tobacco was crossed, first with a variety from the same locality, then with one from the opposite side of the world identical with the first in external appearance. On the other hand, strains of tobacco from seed grown in Connecticut when crossed with plants of the same varieties from seed grown in Italy showed no increase in vigor. Hence, the mere fact of residence in different parts of the world—that is, exposure to different environmental conditions—has no necessary relation to the phenomenon of hybrid vigor, for such individuals may be alike in constitution. Darwin's repeated emphasis of the good derived from crossing plants whose ancestors were exposed to different conditions was because he thought such differences in environment brought about germinal changes. This attitude, therefore, does not detract from his general position that it is differences in germinal construction which bring about hybrid vigor; and this is the principal point at issue.

The manifestations of heterosis are most noticeable as increases in size. This gain in size in plants which are more or less determinate in their number of parts is made up of an increase in the size of parts rather than in the number of parts. In maize the number of nodes is increased much less in comparison to length of internodes. For example, in a large series of crosses between inbred strains of maize height of plant on the average advanced 27 per cent., whereas the number of nodes rose only 6 per cent. Corresponding to the increase in internode length there is an extension in diameter of stalk, length and breadth of leaves. Root development is proportionally augmented. Both the tassels and ears are larger, and frequently two ears develop on crossed plants where either parent produces one, the color of the foliage testifying to the greater vigor.

The greatly enhanced growth of a plant may be made up by increase in the size of cells, as well as by a multiplication in the number of cells. However, in a cross between different species of Catalpa no differences could be seen in tracheid length, although the cross was considerably taller and larger in diameter.

The principal effect of crossing maize is shown by an additional production of seed. A number of crosses have given 180 per cent. increases in yield of grain over their inbred parents. Examples of what can be done are seen in the accompanying illustrations (Figs. 31 and 32). Improvement in yield is shown by crosses between inbred strains derived originally from the same variety, as well as between crosses of strains derived from different varieties or even from quite distinct types. The results have been very wonderful as a whole, giving at the very least a return to the condition of the original stock before inbreeding was commenced. Some combinations regularly give greater increases than others, but in every case such differences are small as compared with those between the crosses and the inbred parents.

Although, in the main, reciprocal crosses give about the same result, some variation in this respect is habitually shown. In general, there is a correlation between the yield of the better parent strain and the yield of the cross. The crosses in which strain No. 1-6 has been used as the female parent have regularly given the highest yields, and this strain is the most vigorous and productive of the four inbred Leaming strains used in our illustrations.

In a comparison of crosses between inbred strains of maize with ordinary outcrossed varieties the inbred hybrids are handicapped because they have to start from small, poorly developed seeds. This handicap is brought out clearly by a comparison of second generation plants grown from self-fertilized seed produced on vigorous hybrid plants, with hybrid plants grown from seed produced on inbred plants. The first generation starts off poorly, as shown in the accompanying illustration (Fig. 33), but soon catches up and passes the second generation. At maturity the second generation is shorter and less productive, although it has a much greater variability. The third generation from selfed plants of this particular cross has been grown, and there is still further loss of the stimulation which is at its maximum in the first generation. On continued inbreeding these families presumably would exhibit a continuation of the same course of reduction in size, vigor and variability shown in the original inbreeding experiment, until homozygosity was again

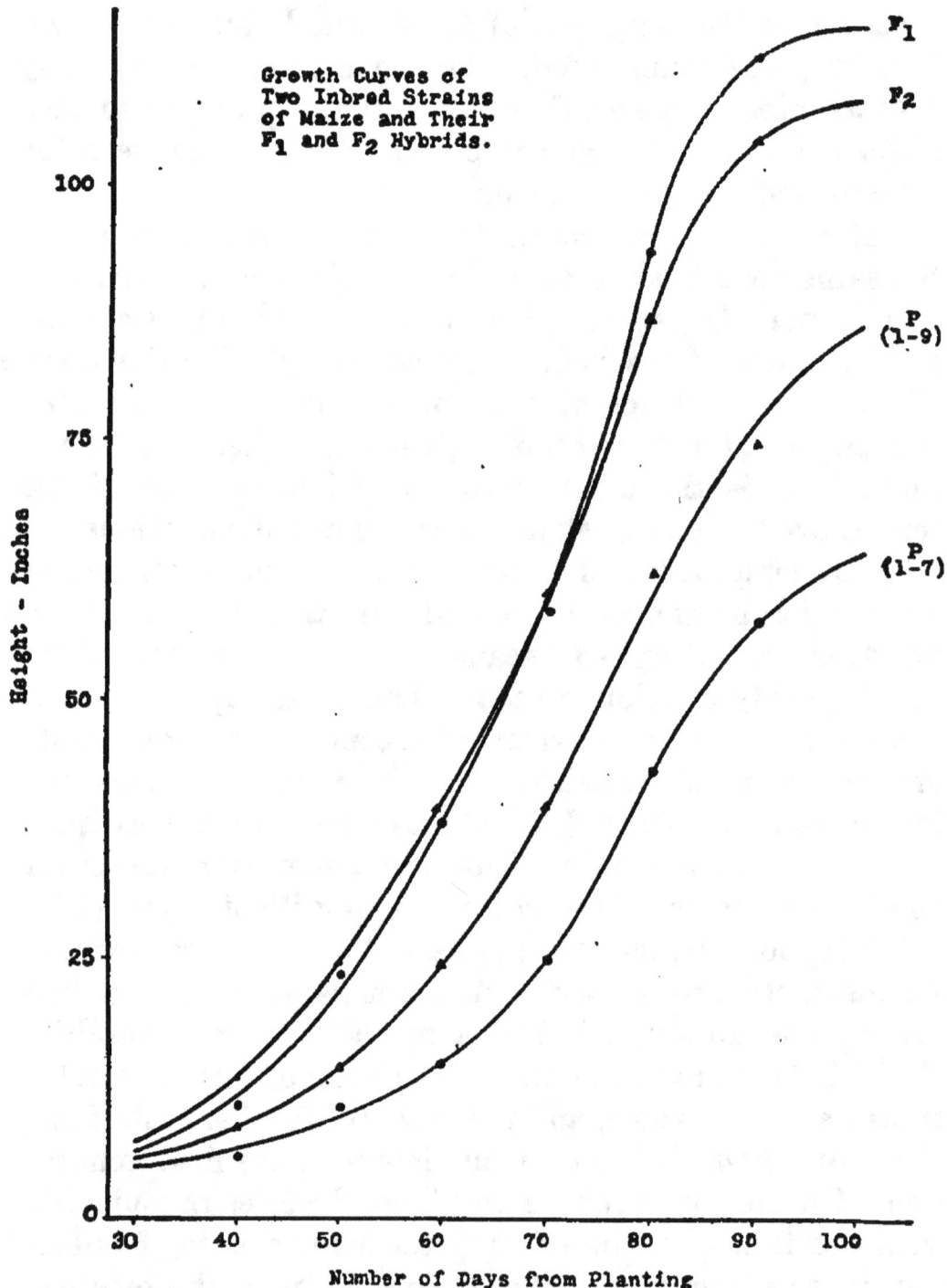

Fig. 33.—Graphs showing growth curves of two inbred strains of maize and their first and second generation hybrids.

reached. The resulting inbred strains would have about the same amount of development as the original inbred strains, but would probably differ from them in appearance through the possession of different combinations of characters. The principal point is that the vigor and size lost by inbreeding are immediately restored by crossing, but lost again on further inbreeding. It is a transitory effect, for the most part, impossible of fixation.

Increases in yield of grain are also frequently obtained when ordinary commercial varieties of maize are crossed. Rarely are the increases greater than 10 per cent., however, and even this is more commonly to be expected when varieties of somewhat different type are used; for example, flint and dent. Most varieties of corn are now so widely crossed and furthermore are so near the limit of production that great advances are not to be expected. Collins [26] has obtained especially large increments in yield by hybridizing types of corn from different geographical regions. Three different varieties of corn from the southwest—Hopi, Brownsville and Hairy Mexican—each gave an increase of 100 per cent. or more when crossed with a variety from China having seeds with a different type of endosperm.

Even before the plants are obtained there is a striking effect of crossing in an immediate increase in the size of seed. This was noted by Roberts,[185] and established very clearly by Collins and Kempton [28] through pollinating ears of maize with a mixture of the plant's own pollen and of a different sort. By taking advantage of the phenomenon of double or "endosperm" fertilization, the experiment was so designed that the outcrossed seeds could be distinguished by differences in endosperm color.

Advances in average weight of seed ranging from 3 to 21 per cent. were obtained. With inbred strains as parents, the increases are even greater, ranging from 5 to 35 per cent. The seeds have a heavier embryo as well as a heavier endosperm, yet curiously enough they mature faster than the selfed seeds on the same ears.

It is a point of some interest, perhaps, that there is no selective action favoring the foreign pollen when these pollen mixtures are applied. This matter has been determined very carefully on account of its bearing on Mendelian theory, but it also answers in the negative the question of whether there is an effect of heterosis manifested by a selective chemotropism before the zygote is formed.

Darwin, in "Cross and Self-Fertilization in the Vegetable Kingdom," compares the time of flowering of 28 crosses between different types of plants which had shown distinct evidence of hybrid vigor. Of them, 81 per cent. flowered before the parents. In other cases, where no heterosis was shown in other characters there was no acceleration of the blooming period. These results have been corroborated in crosses between garden varieties of tomatoes and of sweet corn, where a tendency to put forward the time of both flowering and maturing has been shown to accompany increases in size. Shortening the time of growth thus seems to be one of the many expressions of an increased metabolic efficiency on the part of the hybrid plant.

Increased longevity, viability, endurance against unfavorable climatic conditions, and resistance to disease have also been frequently noted as properties of hybrids. Kölreuter [125] and Wiegmann [70] both mention these points, and Gärtner[74] gives them his especial attention. Under the heading, "Ausdauer und Lebenstenacität der Bastardpflanzen," he makes the following statements:

> There is certainly no essential difference between annual and biennial plants and between these and perennials in regard to their longevity, for frequently different individuals of the same species have a longer life at times as, for example, *Draba verna* which has both annual and biennial forms. The longevity of a plant thereby furnishes no specific difference but at most only signifies a variability. However, in hybrids this difference deserves special consideration. In most hybrids an increased longevity and greater endurance can be observed as compared to their parental races even if they come into bloom a year earlier. The union of an annual, herbaceous female plant with a perennial, shrubby species does not shorten the life cycle of the forthcoming hybrid, as the union of *Hyoscyamus agrestis* with *niger*, *Nicotiana rustica* with *perennis*, *Calceolaria plantaginea* with *rugosa* shows. So also in reciprocal crosses when the perennial species furnishes the seed and the annual species supplies the pollen, as *Nicotiana glauca* with *Langsdorffii*, *Dianthus caryophyllus* with *chinensis;* *Malva sylvestris* with *mauritiana* or biennials with perennials and reciprocally, as *Digitalis purpurea* with *ochroleuca* or *lutea*, and *lutea* with *purpurea*, or *ochroleuca* with *purpurea*. From the union of two races of different longevity a hybrid usually results into which the longer life of one or the other of its parent races is carried whether it comes from the male or female parent species.

Many more instances are given by Gärtner supporting the conclusion previously reached by Kölreuter that the longer life of hybrid plants is to be counted among their usual properties.

Gärtner also gives several examples of endurance to unfavorable weather conditions by hybrids. Many of his tobacco hybrids actually survived the winters in the open field in south Germany when the parents were killed.

The hardiness of hybrids is frequently shown by a great resistance to parasitism. Gernert[77] states that teosinte and the first generation cross of teosinte and maize are not attacked by the aphids which damage maize. The cross between the inbred strain of maize most susceptible to smut, previously mentioned, and the strain not affected, gives a hybrid which is only slightly parasitized. The same thing has been noted in crosses between other strains of maize, some of which are quite badly damaged by an unidentified leaf blight organism. Radish seedlings which were naturally cross-fertilized were much less damaged by damping-off fungus than uncrossed seedlings from the same plants. Resistance is not shown by all first generation hybrids when the parents differ in susceptibility. Some cases are known in which the hybrids are fully as susceptible as the less immune parent. In the majority of crosses reported, however, in which resistance to parasitism is a factor the hybrids tend to show resistance.

Among the diverse manifestations accompanying heterozygosity may be mentioned viability of seed. In maize, crossed and selfed seeds from the same ears have shown a difference of 16 per cent. germination in favor of the crossed seeds. The crossed seedlings appeared earlier and grew faster from the first.

Increased facility of vegetative propagation of hybrids was frequently noted by the early hybridizers. Sageret[191] makes particular note of a hybrid tobacco which easily propagated itself vegetatively. Many of our cultivated fruits which are propagated by buds, grafts, cuttings, etc., owe part of their excellence at least to the fact that they are in a heterozygous condition. Moreover, there is no evidence to prove that plants lose any of their hybrid vigor in long continued vegetative multiplication through innumerable generations.

In general, as noted before, there is similarity between the effect of heterozygosis and that of a good environment. Those characters which are quickest to be modified by external factors also show the greatest change on crossing. A good illustration of this is a Nicotiana cross which was above the average of the parents in both height and leaf size. The length of the corollas, on the other hand, a character very slightly affected by the environment, was not increased. There is at least one difference between the two, however; in time of maturity, environment and heterosis have somewhat opposite effects. Generally speaking, favorable growing conditions tend to delay flowering and maturing, whereas conditions which stunt the plants tend, like heterosis, to hasten them.

Each of these effects is by no means always present when a cross is made. The usual and of course the most noticeable effect is the increase in size. But crossing may have a stimulating effect upon certain parts and a depressing effect on others. This is shown in many species crosses in which reproductive ability is greatly reduced or even totally eliminated, while at the same time vegetative growth is enormously increased. Freeman and Sax have independently obtained seeds from crosses between common bread wheats and macaroni wheats which were shrunken in appearance and small in size, owing to a poor development of the endosperm. The embryos were well developed, however, and the plants produced gave distinct evidence of hybrid vigor.

In this discussion there has been a noticeable omission of the effect of crossing on animals. Illustrations are not lacking that crossing frequently is highly beneficial to animals; but animals do not furnish as desirable research material for this particular problem as plants, on account of their bisexuality, as was explained earlier, and for this reason but few quantitative data are available. There is no question but that animals behave the same as plants in heredity; therefore, one might transfer the conclusions reached in the one kingdom to the other without apology, for the effects of inbreeding and cross-breeding are wholly and solely the working out of the laws of heredity. At the same time, it will not be amiss to present some of the results obtained by zoölogists, for they strengthen the case immensely.

In the two cultivated species of insects which form our sole instances of domestication, bees and silkworms, there seems to be evidence of increased vigor on crossing only in silkworms (Toyama [205]). In the fruit fly, however, upon which the greatest amount of genetic work has been done, Castle,[21] Moenkhaus,[144] Hyde [98] and Muller [154] all found size, fecundity and general constitutional vigor increased remarkably, particularly when the strains crossed had been inbred previously. In the rotifer, Hydatina, Whitney [216] and A. F. Shull [198] obtained similar results. Further, Gerschler [76] describes and figures first generation crosses between different genera of fishes which show very marked increases in size.

In birds also there is such an increase in vigor that poultry fanciers often cross two distinct strains and sell the progeny because of their rapid growth and large size. No attempt is made to breed from the hybrids; they are simply produced because of their vigor. When very great differences in size exist, there is not, of course, an increase in size sufficient to throw the individual of the first hybrid generation above the larger parent, as is shown by the work of Phillips [178] on crosses between the large French Rouen and the small domestic Mallard duck, and by the work of Punnett [182] on crosses betwen the Silver Sebright bantam and Gold-pencilled Hamburgh breeds of poultry. There is an increase over the average of the two parents, but the F_1's do not reach the size of the larger parent race. Part of the reason for the comparatively small sizes of the F_1's in these crosses, however, is due to the fact that the crosses were always made on the *small* hens allowing the hybrid birds to get their start in life with only the nutriment stored in the *smaller* eggs.

The greatest amount of data on this subject, just as there is the greatest amount of interest, has been obtained from the mammals. In the meat breeds of cattle, swine and sheep, as in poultry, it is a common practice to cross distinct races and sell the progeny. The increase in size and the rapidity with which this size is obtained are so general a phenomenon that it bids fair partially to replace the older method of pure line breeding. Not only are varietal crosses thus characterized, but specific crosses. We have already mentioned the mule. With the disadvantage attached to sterility, the mule certainly would not have held its own throughout the past forty centuries were it not for its tremendous capacity for work and its remarkable resistance to disease. Crosses between the ass and the zebra, and between the cow and the zebu also give animals of considerable merit, and one can hardly refrain from thinking that within a few years some considerable use will be made of them.

For precise data on the effect of crossing different races, however, we must turn to the small mammals used so constantly in experimental work, the mouse, the rat, the guinea-pig and the rabbit. One need go no further than to cite the work of Castle and his students at the Bussey Institution of Harvard University, the work of Miss King at the Wistar Institute of Anatomy and that of Wright at the Bureau of Animal Industry of the United States Department of Agriculture. The painstaking researches of these investigators show without question that the effect of crossing on animals is the same as upon plants.

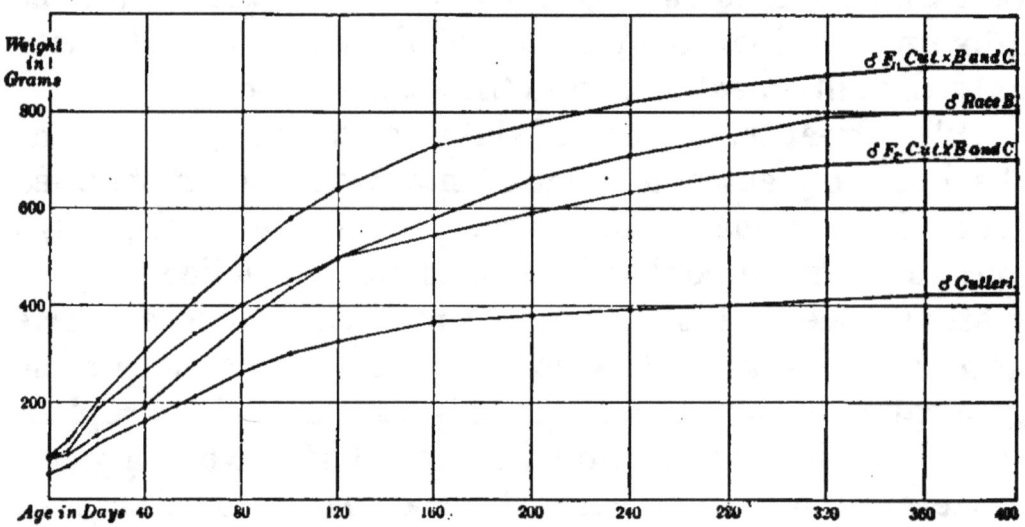

Fig. 35.—Growth curves of males of race B guinea-pigs and Cavia cutleri and their F_1 and F_2 hybrids. (After Castle.)

The results from one genus is typical of them all. Castle [20] made a cross between a domestic guinea-pig and a wild cavy, *Cavia cutleri*. The first generation hybrid males weighed about 85 grams at birth, which is slightly more than the young of either pure race, and retained this lead throughout their subsequent life as is shown by the growth curve in Fig. 35. At maturity they weighed about 890 grams, as compared with 800 grams for the guinea-pig ancestor and 420 grams for the *cutleri* ancestor.

The second generation hybrids of both sexes were

smaller than the first generation hybrids from birth on, showing that some of the growth impetus produced by the hybridization had not been retained. But the growth curve of the second generation hybrids rises rapidly at first, showing the healthy start in life they obtained from their vigorous F_1 mothers.

Perhaps no such increase in vigor as that shown in the species cross just described is usually found when different sub-races are crossed. It would not be expected, for ordinary races of mammals are continually being crossed within the variety and, therefore, hybridization would not be expected to increase heterozygosis to any marked degree. But results similar to those obtained in plants may be expected if the genetic conditions are similar. This is proved by the data Wright obtained when he crossed guinea-pigs born of unrelated inbred mothers and fathers. The cross-breds were distinctly superior to their inbred relatives in nearly all characters connected with vigor. In spite of the fact that their inbred mothers were small and somewhat deficient in vigor, a slightly larger per cent. of cross-breds than of inbreds were born alive, and a distinctly larger per cent. of those born alive were raised. They were somewhat heavier at birth in a given size of litter and gained in weight much more rapidly between birth and weaning. They matured earlier and produced larger litters and produced them more regularly than the inbreds.

Thus the results with animals are comparable to those obtained with plants in all essential features. Briefly, in crosses which are fertile the effects are such as to contribute to a greatly increased reproductive ability, making possible a larger number of offspring. The degree to which heterosis is expressed is correlated, within limits, with the differences in the uniting gametes. When homozygous forms are crossed, it is at its maximum in the first hybrid generation, and diminishes in subsequent generations of inbreeding as segregation occurs and homozygosity is again attained. It is a widespread phenomenon and accompanies heterogeneity of germinal constitution whether the organisms crossed are from the same or diverse stocks, whether they have been produced under similar or under different environmental conditions; although it is not apparent until the zygote is formed, from that time on it is expressed in many ways throughout the lifetime of the individual and is undiminished by asexual propagation.

These are the effects of cross-breeding upon development in which we have been particularly interested, those in which the organizations of the combining gametes are sufficiently compatible to permit continued propagation. But it must not be forgotten that we have dealt with only one part of the problem. As the differences between the forms increase limits are reached beyond which the organisms neither reproduce nor flourish. One can arrange a series in plants in which (1) the parents are so diverse the cross cannot be made; (2) the seed obtained fails to germinate under any set of conditions; (3) the hybrids are so weak they are unable to reach maturity; (4) the hybrids are extremely vigorous, but sterile except possibly in back-crosses; (5) the hybrids are fully fertile and more vigorous than either parent; or (6) the parents may be so closely related no effects whatever are to be noted.

A somewhat similar series can be arranged with animals, although usually in wide crosses if the hybrids can be obtained at all they are as large or larger than the average between parents. A satisfactory interpretation of the vigor of hybridization must take all these facts into consideration, even though they may not be the result of the operation of one single law.

CHAPTER VIII

CONCEPTIONS AS TO THE CAUSE OF HYBRID VIGOR

The early plant hybridizers, although they frequently discussed the increased size and vigor of their crosses, seldom commented on the effect of inbreeding, and made no speculations as to the cause of either. The animal breeders of the period were more imaginative. Acquainted with both phenomena, but more familiar with the results of inbreeding, they unhesitatingly linked the two —the first as an antidote for the second. They attributed most of the injurious effects which appeared in their herds to the concentration of undesirable traits. If unfavorable characters and tendencies to disease were present, mating similar animals brought out these undesirables more pronouncedly; whereas, if healthy animals from unrelated herds were brought in, such tendencies were checked, the defects disappeared, and the health and vigor of the herds returned.

Darwin, however, refused to ascribe any large part of the effects of inbreeding to this cause. He knew of many cases in which *weakened* animals from different inbred herds had been mated together, and gave progeny of full health and vigor and of increased size. The undesirable features induced in both herds by inbreeding disappeared when animals of the different herds were mated. Instead of a concentration of the less favorable traits of the two parental lines the reverse seemed to have occurred. Similar cases in plants were familiar to him, and

CAUSE OF HYBRID VIGOR

proved beyond question the great advantage to be gained by crossing even when the individuals themselves were weak. These facts, taken together with the many marvelous and intricate contrivances of plants to insure cross-pollination, led him to believe that self-fertilization was *inherently* harmful and something to be avoided if possible. The benefits accruing from crossing he ascribed, as we have seen, to the meeting of sexual elements having diverse constitutions.

After Darwin's contribution to the problem of inbreeding no progress was made until less than two decades ago, when the Mendelian discovery opened up so many new possibilities. The conception of an inheritance made up of separable units aroused a new interest in the matter and made possible a unified and satisfactory interpretation of all the facts.

Mendel had shown that characters from one parent might disappear completely in the progeny only to reappear in subsequent generations in some of the offspring. Surely here was something of importance to the inbreeding problem. Unfavorable characters might vanish when different organisms were crossed; but they were merely hidden. Inbreeding revealed them for what they were.

Shortly after Mendel's experiments became known, Bateson crossed two pure white flowered varieties of the sweet pea. Instead of having the white flowers characteristic of two parental races, the hybrid flowers were purple. The wild progenitor of the sweet pea has purple flowers. Here was a case in which crossing brought back previously existing conditions, a return to the wild type characters. This phenomenon had been observed long before this time; in fact, it was so well known it had been given a special name. Atavism, or the reappearance of previously existing characters, was immediately put upon a Mendelian basis by hypothecating two separable factors both of which were necessary for the production of the end result. This was proved by the fact that later white flowered plants were obtained which did not produce color on crossing with either of the original white flowered varieties and therefore lacked both factors. The light increased. Some of the chaotic observations of the earlier hybridizers began to be understood as orderly facts.

Ten years after the rediscovery of Mendel's work a symposium was held by the American Society of Naturalists on the "Genotype Hypothesis," an indication of the growing importance of the ideas associated with the name of Johannsen. The basis of the genotype conception is that individuals which are visibly alike may be germinally unlike—merely an extension of the above Mendelian concepts. Johannsen's contribution was the idea that the unit factors of Mendel are relatively constant and stable in whatever combinations they occur, and that the variability of a constantly cross-fertilized population is largely due to the segregation and recombination of these unmodified factors. When such a heterogeneous population is continuously self-fertilized, homozygosity is ultimately attained, and much of the previous variability disappears. Similar individuals making a homozygous, pure breeding population, are known as a pure line, and while they still vary as affected by different environmental conditions, such variability does not respond to selection, and the average condition is not changed. Although there is still a question as to the degree of stability of the Mendelian unit factor, as there is to the degree of stability of the atom, the principle of the pure line has been firmly established by an ever-increasing body of evidence, and is of the utmost importance in a proper understanding of the facts involved in inbreeding and outbreeding.

The first application of these principles to the problem of inbreeding was made with the results from maize already described. It was shown that self-fertilization automatically brings about homozygosity, and with it a reduction of a great deal of the variability commonly shown. Along with this reduction in variability, certain characters manifest themselves which are more or less unfavorable to the plants' best development. Plants with sterile tassels and sterile ears and plants which lack chorophyll appear, cannot reproduce themselves, and are eliminated. Other characters come to light which do not cause the extinction of the plants directly, but which more or less handicap them in their development; for example, partial chlorophyll deficiency, dwarfness, bifurcated organs, contorted stems and deficient root systems.

All of these characters are shown in small numbers in a cross-pollinated field of maize, but not in sufficient frequency to reduce productiveness seriously. As the result of self-fertilization some of the strains obtained possess certain of these unfavorable characters as regular features. Here, then, is an explanation of part of the injurious effects of inbreeding. Unfavorable characters are segregated out, which reduce the developmental efficiency of the plants which possess them. But if unfavorable characters are concentrated in some lines, favorable characters are concentrated in others. Some have more of the favorable than of the unfavorable, hence some strains resulting from inbreeding are better than others.

But all strains in maize are so greatly reduced by inbreeding that none can be compared in productiveness to the normal cross-pollinated plants. Something besides ordinary segregation must be involved in this well-nigh universal effect of inbreeding.

It was apparent that when germinal heterogeneity was at the maximum the greatest vigor was shown. When this heterogeneity was reduced by inbreeding, vigor was lost. Hence, the fundamental fact that hybrid vigor varies directly with heterozygosity was clearly established. To account for the greater vigor and increased development of hybrids, it was only necessary to postulate that a developmental stimulation was evolved when *different* germ plasms were united. This hypothesis (East and Hayes,[58] Shull [196]) satisfied all the essential facts, and for the first time the effects of inbreeding and cross breeding were clearly understood in their true relation to each other. Inbreeding was not a process of continuous degeneration; it was a process of Mendelian segregation, and its effect was directly related to the number and type of characters existing originally in a heterozygous condition. If unfavorable characters were covered up by favorable characters, inbreeding brought them out whenever a simplification of the germ plasm allowed them to appear. Inbreeding was in effect the isolation of homozygous hereditary complexes from an heterozygous hereditary complex. If the best of these combinations failed to attain the development of the original stock, it was thought to be because they were deprived of a stimulus which only accompanied heterozygosity and which seemingly was impossible to fix.

This hypothesis, by associating all the facts of inbreeding and outbreeding with the phenomena of Mendelian heredity was a great step forward. It went as far as it was possible to go at the time it was devised, and it is capable of interpreting all the facts to-day. But it held some disadvantages. The assumption of a physiological stimulation arising from the interaction of *different* hereditary factors was not altogether satisfactory, for it locked the door on any hope of originating pure strains having as much vigor as first generation hybrids. Fortunately, the development of Mendelian heredity has been such that this part of the hypothesis can be superseded.

The basis for this hypothetical stimulation was seen in the fact that fertilization is usually necessary to start the development of the egg. In most cases, without the union with the sperm, the egg cannot divide and development is prevented. The reaction of the different substances brought together at fertilization stimulates cell division and starts development. This made it reasonable to assume that when the egg and sperm differed in hereditary factors stimulus to development was increased and continued throughout the growth of the resulting organism. According to the view of G. H. Shull and of East, it was the interaction of different elements in the nuclei that produced the stimulation. A. F. Shull,[198] on the other hand, assumed the stimulation to result from the interaction of the new substances brought in by the sperm with the maternal cytoplasm. In his opinion the stimulation might persist for a time even after homozygosis was attained, because foreign elments brought in by the original cross would still remain to react with the cytoplasmic matter. Moreover, it was further assumed that the stimulation might decrease in long-continued asexual propagation through the cytoplasm becoming adjusted to a heterozygous nucleus. This theory was proposed as an interpretation of a reduction in vigor which he had found in parthenogenetically reproduced rotifers. The recent facts, however, are more in accord with the former view because (1) the stimulus actually is lost as homozygosity is attained, and (2) the evidence of vigor being reduced in continued asexual reproduction is not at all conclusive.

The reasons for holding the whole stimulation hypothesis in abeyance at present has developed from the following facts. In 1910, Keeble and Pellew[116] offered a concrete illustration of a purely Mendelian method by which increased growth could result from crossing. They united two varieties of garden peas, which, as grown by them, each ranged from 5 to 6 feet in height. The first generation grown from this cross was from 7 to 8 feet in height, 2 feet taller than either parent, a result comparable to heterosis. The second generation showed segregation into four classes, one class containing plants as tall as the first generation, two classes having plants similar in height to the two parents, and one class made up of dwarfs shorter than either parent. The two classes of medium tall plants, similar in height, were differentiated in the same manner as the parental races; one had thick stems and short internodes, the other had thin stems and long internodes with fewer of them. The number of plants falling into these four classes agreed closely with the expectation for a dihybrid ratio (9:3:3:1) where two factors showing dominance are concerned.

Keeble and Pellew assumed two hereditary factors to be involved: one producing thick stems, the other long internodes. These factors they designated T and L. One of the parental varieties was medium in height, because it possessed one of these factors; *e.g.*, that for thick stems, but lacked the other. Such a plant had the formula *TT ll*. The other variety was of medium height, because it lacked this *T* factor, but possessed the factor for long internodes, and was given the formula *tt LL*. Both of these factors showed dominance over the allelomorphic condition; hence, the first generation of the cross was taller than either parent because both factors were present. Whether or not later investigations have justified this precise interpretation makes no material difference in the discussion here. Taken as it stands, it is a beautiful instance of the way in which complementary action of dominant factors may increase a character in a first generation hybrid over its expression in either parent.

The investigators attempted to generalize from this experiment and to apply a dominance interpretation to the many other cases in which an increase in growth is occasioned. As the matter stood at that time, however, it was impossible to see why recombination of all the dominant factors concerned in the increased growth of the first generation could not readily be obtained, and hence some individuals be produced having maximum size and vigor, yet unaffected by inbreeding because of their homozygous condition. In other words, in generations after the first it ought to be possible to obtain some strains having all the dominant factors and others with all these dominant factors lacking. Any such race could be rendered homozygous; thereafter, self-fertilization would not result in a less vigorous progeny. And while such results may have been obtained in the peas, investigators have not been able to duplicate them in the many other crosses which showed hybrid vigor. Furthermore, not only was the union of such simple factorial combinations inadequate to account for the frequency of the widespread occurrence of heterosis, but there was another seemingly insurmountable objection to the interpretation. It was pointed out that if heterosis were due solely to dominance of independent factors, the distribution of the second generation would be unsymmetrical in respect to those characters in which an increase was shown in the first generation. This criticism has its basis in the familiar fact that Mendelian expectation in the second hybrid generation where there is complete dominance is always an expansion of the form $(3+1)$ to a power represented by the number of factors. Even with partial dominance the criticism holds, although the lack of symmetry is not so marked.

But in the vast amount of data accumulated upon the inheritance of quantitative characters no such tendencies toward asymmetrical distribution in the second generation are evident. In the majority of cases recorded where hybrid vigor is shown in the first generation, the distribution of the individuals fits the symmetrical curve, commonly known as the Curve of Error, remarkably well.

It is evident, therefore, that the objections raised against the hypothesis of dominance as a means of accounting for heterosis, *as outlined by Keeble and Pellew,* are valid. But both these objections to dominance as an interpretation of heterosis were made before the facts of linkage were known. With linkage these criticisms based upon Mendelian expectancies with independent factors do not hold.

Abundant evidence is fast being accumulated to show

that characters are inherited in groups. The different theories accounting for this linkage of factors make no essential difference in the use to which these facts will be put here. It is only necessary to accept as an established fact that characters are thus inherited and that it is these groups of factors which Mendelize. The chromosome view of heredity will be used, as it is the most probable, the most useful, and permits representation in the simplest manner; but adherence to this view is not necessary for our purpose.

The increasing complexity of Mendelism points very strongly to the probability that the important characters of an organism are determined, or at least affected, by factors represented in practically *all* of the chromosomes or linkage groups. This is comprehensible when it is remembered that height or any other size differentiation is only an expression of an organism's power to develop. Hereditary factors which affect any part of the organism may indirectly determine the maximum of any size character. For example, in plants height is governed by root development as well as by that of the aerial parts.

The widespread occurrence of abnormalities and characters which are detrimental to an organism's best development are well known. It may be taken for granted, nevertheless, that no one individual has all the unfavorable characters, nor, on the other hand, all the favorable characters known to occur in the species. For the most part, each possesses a random sample of the good and the bad. This being true, it is only necessary to assume that in general the favorable characters are in some degree dominant over the unfavorable, and the normal over the abnormal in order to have a reasonable explanation of the increased development of hybrids in the first generation over the average of the parents or subsequent generations. In the first hybrid generation the *maximum* number of *different* factors can be accumulated in any one individual; and because of factor linkage it is extemely difficult to recombine in one organism in later generations any greater number of homozygous characters than were present in the parents, provided the factors are distributed at random in all of the chromosome pairs. This view of the situation makes more understandable why the effects of heterozygosis result in an increase in development, and why they remain throughout the life of the sporophyte, even though innumerable asexual generations.

The abstract view of the dominance hypothesis may be somewhat clearer if a concrete diagrammatic illustration is made. A case will be assumed, in which two homozygous individuals, having three chromosome pairs, both attain the same development as represented by any measurable character. This development will be considered to amount to 6 units, 2 of which are contributed by each chromosome pair. One of these individuals, which we will call "X," attains its development through the operation of factors distributed in the three pairs of chromosomes, each differing from the others in its contribution. Any number of factors can be chosen, but, for the sake of simplicity, only three in each chromosome will be employed. These are numbered 1, 3, 5; 7, 9, 11; and 13, 15, 17 in the accompanying diagram (Fig. 36). The second individual, "Y," develops to an equal extent in the character measured. It attains this same development, however, by the operation of a different set of factors distributed in the three chromosomes and numbered 2, 4, 6; 8, 10, 12; and 14, 16, 18 in the diagram. Both individuals are homo-

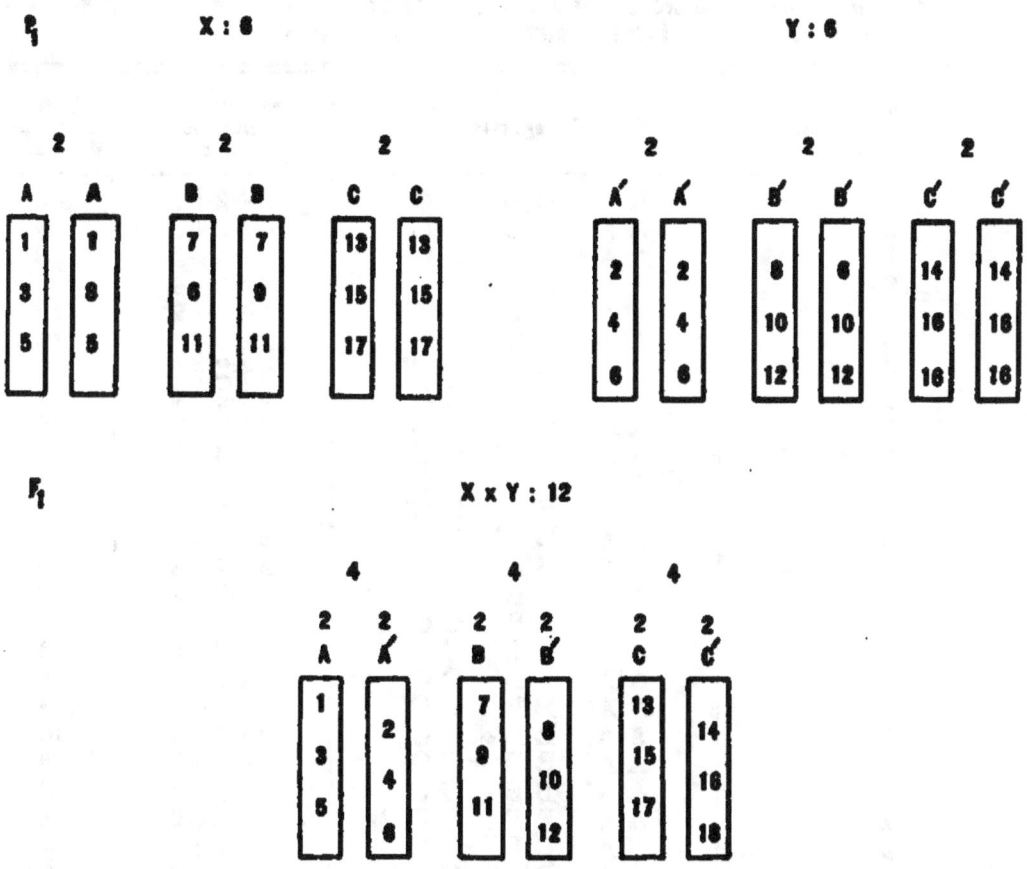

Fig. 36.—To show how factors contributed by each parent may enable the first generation of a cross to obtain a greater development than either parent.

zygous; *i.e.*, the allelomorphic pairs are composed of like elements. It is also assumed that all these nine factors are as fully effective in the haploid as in the diploid condition; in other words, they show perfect dominance over their absence. It will be seen from the diagram that when these individuals are crossed together the progeny develop to twice the extent of either parent, because there are present eighteen different factors instead of nine.

TABLE VII

Composition of a Mendelian Tri-Hybrid in F$_2$ Where the Development Which Each Individual Attains Depends upon the Number of Heterozygous Chromosomes Contained and Thereby upon the Total Number of Different Factors Present.

Number of individuals in each category	Categories	Contributions of each chromosome pair	Total development
1	A A B B C C	2+2+2	6
2	A A' B B C C	4+2+2	8
2	A A B B' C C	2+4+2	8
2	A A B B C C'	2+2+4	8
4	A A' B B' C C	4+4+2	10
4	A A B B' C C'	2+4+4	10
4	A A' B B C C'	4+2+4	10
8	A A' B B' C C'	4+4+4	12
1	A A B B C'C'	2+2+2	6
2	A A B B' C'C'	2+4+2	8
2	A A' B B C'C'	4+2+2	8
4	A A' B B' C'C'	4+4+2	10
1	A A B'B' C C	2+2+2	6
2	A A B'B' C C'	2+2+4	8
2	A A' B'B' C C	4+2+2	8
4	A A' B'B' C C'	4+2+4	10
1	A'A' B B C C	2+2+2	6
2	A'A' B B' C C	2+4+2	8
2	A'A' B B C C'	2+2+4	8
4	A'A' B B' C C'	2+4+4	10
1	A'A' B'B' C C	2+2+2	6
2	A'A' B'B' C C'	2+2+4	8
1	A'A' B B C'C'	2+2+2	6
2	A'A' B B' C'C'	2+4+2	8
1	A A B'B' C'C'	2+2+2	6
2	A A' B'B' C'C'	4+2+2	8
1	A'A' B'B' C'C'	2+2+2	6
64 Total			

Distribution of the F$_2$ individuals according to the development attained

Classes............	6	8	10	12	= 4	Number of classes
Frequency..........	8	24	24	8	=64	Total population

Following this hypothetical case into the second generation by selfing or by interbreeding the individuals of the first generation, the data given in Table VII are obtained. Summing up the results of this tabulation, it will be found that eight individuals are completely homozygous and reach the same development as either parent, six units; eight are heterozygous in all three chromosome pairs and duplicate the twelve-unit growth of the first generation; the remaining forty-eight individuals fall into equal-sized groups, developing to eight and ten units, respectively. In other words, the distribution is symmetrical, and this symmetry remains, however many chromosomes are involved.

It should also be noted that the mean development of the second generation is nine units, which is an excess of just half of the excess of the first generation over the parent. The extra growth derived by crossing the two different types has diminished 50 per cent. In the third generation, from a representative sample of the second generation, it can be shown that this excess again diminishes 50 per cent., so that the effect on the average is only 25 per cent. as great in this generation as in the first, and so on, in subsequent generations, until the effect diminishes to a negligible quantity in about the eighth generation. This is in fair agreement with the actual results obtained by inbreeding maize, as it ought to be, because the development attained by each individual varies directly with the number of heterozygous factors.

In the preceding illustration of the way heterosis may be brought about, perfect dominance was assumed. Moreover, breaks in linkage with the formation of new linkage groups were not considered. All these things enter as complicating factors. Perfect dominance, except in more or less superficial characters, rarely occurs, and even when it does occur, it may be merely an appearance rather than a reality. The general consensus of opinion at the present time is that there is no such thing as perfect dominance, that the heterozygote merely approaches the condition of one or the other parent more or less closely. When two different potentialities are contributed by the parents, there results an interaction between them and the end product is represented in the organism. Because the most striking effect may resemble the character of one parent more than the other, we say that this character is dominant. In reality, in the more fundamental characters, the hybrid usually shows a resemblance to both parents. The more common illustrations of dominance, such as fur colors and flower colors, probably have little to do with heterosis. Other dominant characters, however, have a fundamental effect upon development, nearly always being essential to greatest vigor. Various grades of albinism are common in maize and in many other plants. Since this affects the amount of chlorophyll, the presence of albinism in any form seriously retards the growth. In extreme cases the plants are totally incapable of continuing existence beyond the stage made possible by food stored in the seed. In animals, albinism does not have the physiological significance that it has in plants, but even here it is sometimes unfavorable to the individuals showing it. In every case, and in all degrees, true albinism is recessive to the normal condition. In maize, the heterozygous green plants cannot be distinguished from homozygous green plants. Many other unfavorable characters in maize are also recessive. Absence of brace roots, bifurcated ears, dwarfism, susceptibility to smut all behave in this way.

Certain factors have even been recognized, and in the case of Drosophila [155, 156] have been located in the chromosome mechanism, which are so injurious that they cause the death of the individuals possessing them, unless protected by the factors being in combination with a normal allelomorph. A well-known case of this kind is the yellow mouse. Lethal factors, in order to be recognized easily, must be recessive in their lethal action and must show a visible effect on the soma when in combination with their allelomorphs, since only in that case can the heterozygotes be detected. In the yellow mouse there is associated with color another effect which causes the death of the animals when they are pure for that factor. This has been demonstrated by the altered ratios obtained. Yellow mice are mated together; instead of getting a ratio of 3 yellow to 1 non-yellow, the ratio is more nearly 2:1; that is, (1):2:1, in which the pure recessives are eliminated.[19] This assumption is further corroborated by actually finding the missing number of animals in stages of dissolution in early embryonic life. Of the more than one hundred and twenty-five mutations which have been described in Drosophila by far the greater majority of them are recessive, and nearly all of them are less favorable to the development of the fly than the wild-type characters. The effect of the recessive factors even seems to be cumulative, because when many of them are combined together the flies are extremely difficult to maintain. Much the same condition is true for domesticated animals and plants. The majority of the variations which have occurred are recessive, and are seldom beneficial and often deleterious.

One may be led to inquire why it is that most of the experimentally observed mutations are recessive and less favorable to the best development of the organism. We do not know, but we may hazard a guess. The repeated appearance and disappearance of certain mutations is merely a type of variability which has probably been a constant feature of the organism for a long period and has been subjected to natural selection in the same way as any other character. In other words, may not the tendency to produce dominant unfavorable variations have been reduced to the minimum by natural selection? Conversely, a tendency to produce unfavorable recessive mutations has been tolerated because the latter are protected in hybrid combinations by their dominant favorable allelomorphs. Whether this be true or not, there certainly is a strong tendency for dominant unfavorable variations to be eliminated, because they are constantly subjected to natural selection; while dominant favorable variations, whenever they occur, replace former characters, and become part of the stock in trade of the organism. Recessive mutations, on the other hand, whether favorable or unfavorable, cannot compete for place with natural selection as the judge unless the proper mating brings them into the homozygous condition. If through continuous cross breeding this does not occur, they may be carried on for countless generations—family skeletons hidden by the phenomenon of dominance.

The relation of these reflections to heterosis is just this: If any individual is deficient and handicapped in its hereditary make-up, there is a good chance that this deficiency will be supplied when it is crossed with other individuals, because all are not apt to be wanting in the same things. What one lacks is supplied by the other and conversely. In other words, there is a pooling of hereditary resources, so that the combined effect is better than either could produce alone.

This complementary action can be illustrated by assuming three linked factors, all of which are essential for best development. In one organism there is AbC; in another there is aBc. Dominance, either partial or complete, is characteristic of each. Now in the former interpretation of heterosis, where a physiological stimulation was assumed, the heterozygous combination, Aa, for example, evolved developmental energy and differed in that respect from either AA or aa. Moreover, this stimulation was considered to be of a general nature, affecting the organism as a whole, and was thus differentiated from the specific effect which each had as hereditary factors. With linkage, one may consider heterosis to be due to the action of heredity alone: the hybrid union Aa is not superior to either of the homozygous combinations, AA or aa, but is more or less intermediate. This view has a very great theoretical and practical importance, because one may expect to obtain homozygous instead of heterozygous combinations of the factors which bring about increased vigor in crosses and thus obtain individuals which will have a vigor equal or even superior to the first cross and which will not be affected by future inbreeding.

Such a happy result was not possible on the stimulation hypothesis. This hypothesis was invented to account for the frequency of heterosis, the loss of vigor due to inbreeding, and the extreme rarity of homozygous combinations approaching those of the first hybrid generation in vigor. With a knowledge of independent Mendelian heredity only it was necessary. But if in our illustration the individual AbC-aBc is vigorous because of heredity alone, and if it usually segregates germ cells of the types *AbC* and *aBc,* making the vigor thus obtained a practical corollary of heterozygosity, there is still the chance, no matter how closely linked these factors may be, of breaks occurring which will bring about the production of a gamete *ABC*. This gamete, if it meets another of the same type, will result in a homozygous individual, and if dominance is but partial, this individual, through the very fact of its homozygous condition, will be even more vigorous than those of the first hybrid generation.

Practically the difficulties in the way of obtaining such pure combinations may be very great or even insurmountable, but the hypothesis holds out the hope of thus obtaining types of great economic value. The rearrangement of factors in all possible recombinations is not prevented by linkage as long as there are breaks in the linkage. But since these breaks occur with varied frequency between different factor loci, and in some cases are very rare, the problem is exceedingly complicated; and when many linked factors are involved the chance of obtaining an individual which is completely homozygous in all factors in the first segregating generation is so extremely small that for all practical purposes it may be disregarded. In later generations the chance may become somewhat greater because of the formation of new linked groups. If these are combinations which are favorable to development, natural selection will increase individuals possessing them at the expense of those having less favorable combinations. In time, there is the possibility, however remote, of all the more favorable factors being brought together in a homozygous combination which, therefore, will not be reduced by inbreeding.

Our hypothetical illustration of characters contributed by both parents may be supported by actual results from a cross between inbred strains of maize. As mentioned before, maize strains have been obtained which lack brace roots and are unable to stand upright when the plants become heavy. These strains, however, have the habit of branching freely from the base of the plant, thus producing several main stalks from each seed. When this strain is crossed with one which possesses well-developed roots, but which does not branch at the base of the plant, the result is an extremely vigorous progeny which produces several stalks from each seed and which shows no deficiency in root development. The hybrid plants are so large and so exceedingly vigorous that other factors must have been involved, but these two characters can easily be seen to have contributed something to the luxuriant development.

An even more striking illustration was obtained by Emerson. A dwarf race of maize which was almost completely sterile was crossed with a tall plant which was so deficient in chlorophyll production that it was unable to produce seed, although it had some functional pollen. The hybrid plants were tall, dark green and produced well-developed ears. Here normal stature was contributed by one parent and proper chlorophyll development by the other, the progeny was thereby enabled to develop well and to become highly productive.

In both these crosses the characters involved are largely of a superficial nature, although most of them are necessary for full development. They are characters which are easily seen and serve as an indication of what must have occurred in the case of other factors more internal in their effect and of more fundamental importance. These more fundamental factors are those concerned directly with metabolism and cell division. As to their nature and the way in which they are inherited, we, as yet, know little, but there is reason for supposing they are Mendelian in mode of inheritance and operate in a way to enable the hybrid progeny to attain a greater development than either parent.

For a definite detailed case showing exactly how dominance and linkage thus work together we must look to the work with *Drosophila melanogaster,* as this is the only material which at the present time has been sufficiently well analyzed for our purpose. Bridges and Sturtevant have discovered, isolated, and determined the linkage relations of nearly one hundred factors distributed throughout the four chromosomes of this little fly, in the great majority of which the recessive condition is unfavorable. Through this indefatigable work there is an enormous amount of data from which to choose; but in order to make our illustration comparatively easy to follow, let us consider only four characters which are linked together in the second chromosome. The factors which have the principal effect on these characters may be given the name of the character. They are long legs (D) dominant to "dachs" legs (d), gray body (B) dominant to black body (b), red eye (P) dominant to purple eye (p), and normal wings (V) dominant to vestigial wings (v). In gamete formation in the female there are breaks with a frequency of 10 per cent. in the linkage between d and b, of 6 per cent. between b and p, and of 13 per cent. between p and v, disregarding some disturbing conditions which need not concern us here. In the male there are no linkage breaks.

Now if a female fly with dachs legs, gray body, purple eyes and normal wings ($dBpV$), be crossed with a male having long legs, black body, red eyes and vestigial wings ($DbPv$), the resulting progeny will have the usual wild-type characters, long legs, gray body, red eyes and normal wings, and will be considerably more vigorous than either parent. If these factors segregated independently, one would expect to find one gamete out of every sixteen to be of the constitution $DBPV$, and would obtain one F_2 individual homozygous for this combination of the four dominants out of every two hundred and fifty-six. As a matter of fact, owing to the linkage relations found, only one gamete of this kind is produced in two thousand and then only in the female. It is, therefore, impossible to obtain the type sought in the F_2 generation. But males of the all-dominant type will appear in F_2, and the pure strain may be established in F_3. The word "may" is used as a sort of forlorn hope, however. There *is* a possibility of establishing the homozygous dominant strain in F_3, but when one realizes that in F_2 only one such male and one heterozygous female similar in appearance to hundreds of her sisters will be produced in every four thousand progeny, the difficulties in the situation are emphasized.

The frequency of the linkage breaks is large and the number of factors small in this illustration. When it is remembered that in other organisms there are ten, twenty, or even forty chromosome pairs to be considered, with possibly dozens of factor differences, instead of four in each chromosome, some idea may be obtained of the real difficulties involved in producing individuals of maximum vigor unaffected by inbreeding. Practically speaking, it is impossible unless dealing with a small number of loosely linked factors, except when long periods of time are available and when natural elimination of undesirables is high.

In tracing the evolution of ideas concerning the effects of inbreeding and outbreeding we must give great credit to Darwin for calling attention to the importance of the phenomena in relation to evolution and for being the first to see that hereditary differences, rather than the mere act of crossing, was the real point involved; but with all due credit to Darwin, it was not until Mendelism became known, appreciated and applied that the first real attack upon the problem was made possible. When linked with Mendelian phenomena it was clearly recognized for the first time that one and the same principle was involved in the effects of inbreeding and the directly opposite effects of outbreeding. Inbreeding was not a process of continual degeneration. Injurious effects, if present, were due to the segregation of characters. In addition to this segregation of characters the fact was established that an increased growth accompanied the heterozygous condition. All the essential facts were accounted for. A decade later the great extension of knowledge in the field of heredity has made possible a still closer linking of the facts of inbreeding and outbreeding with Mendelism. The hypothesis of the complementary action of dominant factors is the logical outgrowth of former views and makes the increased growth of hybrids somewhat more understandable. The fact of a stimulation accompanying heterozygosity is supplemented by a reason why such an effect is obtained. The former view of a physiological stimulation and the more recent conception of the combined action of dominant factors are not then two unrelated hypotheses to be held up for the choosing of the one from the other. The outstanding feature of the latter view is that there is no longer any question as to whether or not inbreeding as a process in itself is injurious. Homozygosity, when obtained with the combination of all the most favorable characters, is the most effective condition for the purpose of growth and reproduction.

CHAPTER IX

STERILITY AND ITS RELATION TO INBREEDING AND CROSS-BREEDING

PROBABLY the most noticeable effect of inbreeding in both animals and plants is a reduction in fertility in the earlier inbred generations. The experiments of Ritzema-Bos [184] with rats, of Weismann [86] with mice and of Wright [225] with guinea-pigs are all thus characterized.

Miss King,[119] on the other hand, has inbred albino rats for twenty-five successive generations by brother and sister mating without any appreciable reduction in fertility. Similarly, Castle [21] and his students maintained the fertility of Drosophila for fifty-nine generations of brother and sister mating by breeding from the most fertile flies. Various lines were isolated, nevertheless, which differed in the number of offspring produced, and in the first part of the experiment many individuals appeared which were absolutely sterile. The production of such non-fertile flies became less in the latter part of the experiment, and the average fertility of the remaining stock was improved by this elimination.

In maize the results of inbreeding are generally quite serious as regards fertility. In the first place the consistent reduction in size and constitutional vigor of the plants necessitates a much smaller production of pollen and ovules. The tassels are reduced in size and have fewer branches. The ears are smaller and shorter and oftentimes imperfectly covered with seeds even when abundant pollen is available. In some cases the leaves enclosing the tassels do not unfold properly and the tassel does not develop as it should. This is a secondary effect, but, nevertheless, is one factor in reducing fertility. The anthers are frequently much shrunken, sometimes shedding no pollen at all and even under the best conditions producing a very meagre supply. The amount of pollen produced is more affected by weather conditions in such inbred strains than in more vigorous plants. At the same time inbred strains of maize have been obtained which show no degeneration in the staminate parts. Their anthers are full and produce abundant pollen. Such strains thus far have been few in number. They are correlated with poor development of the pistillate parts. Those strains which have the best developed ears as a rule have very much reduced tassels with a large amount of pollen abortive. Some strains have been obtained which are about equally well developed in both staminate and pistillate functions, and these range all the way from plants which are fairly productive for inbred strains down to types which barely produce enough seed to survive, and since many cases of failure to produce seed are met there can hardly be a doubt that in some of them a complete abortion of one or both functions has taken place. As in the many other effects of inbreeding different results are produced in different lines, showing clearly that segregation of certain factors influencing fertility has taken place. On the whole, there is in this species a tendency for inbreeding to result in a change from a monœcious condition to a functionally diœcious condition.

Sterility in the form of structural degeneration when it occurs gradually increases upon inbreeding until homozygosity is attained, but for the most part it does not show any clear-cut segregation. Yet reduction in fertility is noticeable only so long as there is a change in other characters, constancy in visible characters being accompanied by a constancy in the matter of fertility. In other words, there is no more an accumulation of sterility on continued inbreeding than there is an accumulation of any other effect. Any reduction in fertility ceases when homozygosity is reached, but the end result may be decidedly different in various lines coming originally from the same stock.

Many other instances of an effect of inbreeding upon fertility might be given, particularly the appearance of abnormalities in the genital organs, both external and internal. But what we desire is to show their meaning rather than to catalogue them, and for this purpose no data have been gathered as valuable as those upon the much cited maize. Examination of all the isolated facts brought to light in both animals and plants shows such a similar trend that there is no reason to believe we are not dealing with manifestations of one and the same law, yet only in this species do we have a critical test of the hypotheses involved. And here it can be stated unequivocally and without reservation that the effect of inbreeding on fertility is exactly the same as its effect upon other characters. Recessive combinations deleterious to the function of reproduction are brought to light. But this is not the only conclusion to be drawn. The frequency with which depression of fertility occurs during inbreeding, the slowness with which it is brought to an end, the variety of differences which is brought out, all show how complex one must conclude the function of reproduction to be, and how many variations affecting it must be constantly occurring. In other words, there is here a concrete illustration of the primary importance of reproduction in all evolution. Since provision for succession exceeds all other matters in import to the species, new variations are constantly taking place, new processes are continuously being tried out. The result is to have reproduction tied up with more complications than any other physiological process, to have in naturally cross-bred species more heterozygous factors than in any other character complex. Reproduction, therefore, as we have seen, is affected more often and more frequently than anything else when inbreeding occurs.

When inbred strains showing reduced fertility are crossed, on the other hand, there is almost always a return to the original productiveness along with the return to the original size and vigor. In fact, just as fertility is affected adversely by inbreeding more than any other character, so is it increased more in proportion by crossing. But such increases in productiveness are the rule only up to a certain point of germinal difference between the individuals taking part in the cross. As dissimilarity in the uniting germ plasms becomes greater, sterility again manifests itself. This time, however, the sterility shown is of a different nature. No structural abnormalities appear. There are no variations such as are found in the numerous strains differentiated by inbreeding. It is simply a matter of non-production of functional gametes.

Based upon these germinal differences crosses between species may be classified arbitrarily as follows:

(1) The hybrid may have the same or greater vigor and fertility than the parents. *Nicotiana alata* and *N. Langsdorffii*, for example, are distinct species having dif-

ferences in many characters, yet their hybrids give no indication of any lessened fertility.

(2) The hybrid may have the same or greater vigor than the parents and at the same time show reduced fertility or even total sterility. This is a common result with many species hybrids. The increase in growth is oftentimes extreme. The cross between the garden radish, *Raphanus sativus,* and the cabbage, *Brassica oleracea,* two species belonging to different genera, gives plants of rampant growth which are very nearly, if not completely, sterile, as shown by Sageret [191] nearly a century ago and by Gravatt [85] in recent times (Fig. 38). In most of these cases no seed can be produced by self-fertilization, but back crossing with one or the other parents is sometimes successful. Animal hybrids frequently show sterility in the males and partial or complete fertility in the females. This is the condition in Cavia species hybrids (Detlefsen [47]) and in crosses made between the buffalo, *Bison americanus;* the yak, *Bibos gruniens;* the gayal, *Bibos frontalis;* the gaur, *Bibos gaurus,* and the domestic cow, *Bos taurus.*

(3) The species hybrid may exhibit a reduced size and a decline in vigor combined with complete sterility. The phenomenon shows in various degrees. For example, East and Hayes [59] made several *Nicotiana* crosses in which the seed would not germinate, although both embryo and endosperm tissue was formed. Crosses between *Nicotiana tabacum* and *N. paniculata,* and between *N. rustica* and *N. alata* resulted in seed which germinated, but the plants were weak and died before flowering, apparently because of inability to utilize the starch formed.

In other crosses the plants matured, but they developed very slowly and in the end were smaller than either of the parents.

In general, therefore, it can be said that differences in uniting germ plasms, when not too great, may bring about both more efficient development and increased fertility. Beyond that critical point of difference both fertility and vigor may be decreased, but fertility is usually the first to suffer—even complete sterility often being coupled with rampant growth. Nature thus steps in before a germinal heterogeneity which will endanger the health of the hybrid organism has been reached, and prevents multiplication entirely. This is an important physiological provision, since when great germinal differences exist there is reduced growth as well as sterility. Groups are thus set apart which may evolve within themselves by putting to good use heterosis and Mendelian recombination. What apparently happens is this: As germinal differences increase a point is reached at which the precise and complex machinery governing gametogenesis cannot do its work in the normal manner and sterility results, although under the same conditions developmental cell division goes on as usual. Beyond this degree of difference in the uniting germ plasms, even somatic cell division is affected.

This sterility accompanying wide crosses is an almost untouched problem. We can throw no light upon it except the suggestions noted in the last few sentences. For this reason one may inquire why it is mentioned in this connection at all. In spite of our comparative lack of knowledge as to just what occurs in the cell divisions of wide crosses, however, there is an excuse for meddling. The peculiar resemblance of the effect of inbreeding to the effect of crossing various distinct species, has led many writers to identify the phenomena. Further, several critics have maintained that a theory which purports to interpret sterility in the one case, should interpret it in the other. Now this is a point of view which is obviously incorrect even with our present meagre knowledge of the facts. The sterility often accompanying inbreeding is not the same thing as the sterility resulting from hybridization. The resemblance is superficial in the extreme. In the one case there is the differentiation of distinct strains differing anatomically and physiologically in their ability to perform the act of reproduction. It is a phenomenon of Mendelian heredity which stands out in the clear-cut manner it does, *because the progenitors of the individuals thus characterized have gone through with the mechanical process which segregates factors, in the precise manner necessary to accomplish the purpose. In the other case, the individuals are sterile because they cannot go through this same process in the exact and proper way required, on account of the incompatibility of the uniting cells.*

CHAPTER X

THE RÔLE OF INBREEDING AND OUTBREEDING IN EVOLUTION

In our brief consideration of the more important changes which have occurred in the reproductive mechanisms of animals and plants, several features stand out impressively. Both animals and plants have followed modes of reproduction that are identical in what are deemed to be the essential features, something which can be said of no other life process. It is not enough simply to say that sexual reproduction has become the dominant mode of propagation among organisms. One must go further. Cross-fertilization, either continuous or occasional, is the really successful method of multiplication everywhere. Such a parallel evolution in the two kingdoms is valid evidence of real worth in the process: a consideration of the evidence on inbreeding and crossbreeding permits us to state this value in concrete terms.

The establishment of methods of reproduction which maintain variation and inheritance mechanisms on a high plane of efficiency is naturally a fundamental requirement in evolution. Since, however, we have seen that there is no reason for believing sexual reproduction to be better adapted to assure a numerous progeny than asexual reproduction, it either must be a more perfect means of hereditary transmission, or it must offer selective agencies a greater variety of raw material.

Fortunately we are able to eliminate the first alternative. There is definite evidence that sexual reproduction does not differ from asexual reproduction in what may be called the *heredity coefficient*. It holds out no advantage as an actual means for the transmission of characters.

The majority of zoölogical data on this subject has little value on account of the experimental difficulties inherent in the material, although zoölogists have published more on the matter than the botanists. Plants furnish the best material because of the ease in handling large numbers of both cuttings and seedlings side by side, and because of the opportunity to utilize hermaphroditic species. Even with the best plant material several undesired variables are present, and experiments with them, therefore, are not without their disappointments; but no one who has had a long and intimate experience in handling pedigree cultures of plants can have any doubts concerning the correctness of our conclusion. Practically the inquiry must take the form of a comparison between the variability of a homozygous race when propagated by seeds and when propagated by some asexual method. The first difficulty is that of obtaining a homozygous race and thus eliminating Mendelian recombination. The traditionally greater variability of seed-propagated strains is due wholly to this difficulty, we believe. It may be impossible to obtain a race homozygous in all factors. There may be a physiological limit to homozygosis even in hermaphroditic plants. The best one can do is to use a species which is naturally self-fertilized, relying on continued self-fertilization for the elimination of all the heterozygous characters possible. We have examined many populations of this character in the genus Nicotiana and have been astounded at the extremely narrow variability they exhibit. Even though one cannot grow each member of such a population under identical conditions as to nutrition, the plants impress one as if each had been cut out with the same die. Qualitative characters such as color show no greater variation, as far as human vision may determine, than descendants of the same mother plant propagated by cuttings. Further, in certain characters affected but slightly by external conditions, such as flower size, the sexually produced population not only shows no greater variability than the asexually produced population, but it shows no more than is displayed by a single plant. Yet one must remember that in such a test the seeds necessarily contain but a small quantity of nutrients and for this reason the individual plants are produced under somewhat more varied conditions than those resulting from cuttings, hence it would not have been unreasonable to have predicted a slightly greater variability for the sexually produced population, even though the coefficient of heredity of both were the same. Similar, though less systematic, observations have been made on wheat—an autogamous plant almost as satisfactory for such a test as Nicotiana—with practically identical results.

One is justified, then, in claiming there is experimental evidence to show that sexual reproduction in itself is no more than an exact equivalent of asexual reproduction in the matter of an heredity coefficient. But is this also true for germinal variation? We believe it is. Variations similar in size and kind arise both in asexual and in sexual reproduction, but it cannot be maintained they occur more frequently in the latter. There are insects in Oligocene amber apparently identical with those of to-day, proving constancy of type to be possible under sexual reproduction through millions of years; there are asexually reproduced species of plants just as constant and probably still more ancient. At the same time, germinal variations occur to-day under sexual reproduction in somewhat noteworthy numbers, as Morgan's work on Drosophila shows. There has been no trustworthy estimation of their frequency within even a single species, but it cannot be said they occur in less numbers than where asexual reproduction rules, even among organisms of a relatively high specialization. If there are those who doubt this statement, let them refer to the huge list of bud-variations in the higher plants compiled by Cramer.[33] He will be able to identify there, type by type, class by class, practically all of the variations he is able to discover in the same species in the literature on seminal reproduction.

It is rather odd that this should be the case, for what is being discussed here is not really the frequency with which variations occur, but rather the frequency with which they are detected. And theoretically, the ease of detecting variations ought not to be the same under the different modes of reproduction. If it be granted that changes in the constitution of the chromosomes are direct causes of variations, and that such changes in constitution are equally probable in all chromosomes, it follows that *parthenogenetic* individuals having the *haploid* number of chromosomes should show a larger proportion of germinal variations than members of the same species having the diploid number of chromosomes, because variations of all kinds would be recognizable in the former case, while in the latter recessive variations could not be detected until the first or second filial generation and then only when the proper mating was made. Though there is no direct support for this idea in the species where the premises hold, there is some evidence that the reasoning is not wholly improbable. Bud-variations occur much more frequently in heterozygotes than in homozygotes. This simply means that bud-variations are brought to light more frequently in heterozygotes than in homozygotes: and a reason is not hard to find. Recessive variations are much more frequent than dominant variations, and a recessive variation in a particular character shows only when the organism is heterozygous for that character. If a recessive bud-variation arises in a homozygote and gametes are afterwards developed from the sporting branch, it is not at all unlikely that the variation may show in the next generation, but it will be attributed then to gametic mutation.

If, therefore, one is constrained to admit that the preponderance of the evidence points to practically the same coefficient of heredity for both forms of reproduction, and that variation in the sense of actual changes in germinal constitution *may* occur with greater frequency in asexual reproduction, if there be any difference at all between the two forms, he is left with only one reasonable hypothesis to account for everything, *Mendelian segregation and recombination.*

Mendelian heredity is a manifestation of sexual reproduction. Wherever it occurs, there Mendelian heredity will be found. Now if N variations occur in the germplasm of an asexually reproducing organism, only N types can be formed to offer raw material to selective agencies. But if N variations occur in the germ plasm of a sexually reproducing organism 2^n types can be formed. The advantage is almost incalculable. Ten variations in an asexual species mean simply 10 types; 10 variations in a sexual species mean the possibility of 1024 types. Twenty variations in the one case is again only 20 types to survive or perish in the struggle for existence; 20 variations, in the other case, may present 1,032,576 types to compete in the struggle. It is necessary to condition the argument by pointing out that these figures are the maximum possibilities in favor of sexual reproduction. It is improbable that they ever actually occur in nature, for 2^{20} types really to be found in the wild competing for place after only 20 germinal variations would mean an enormous number of individuals even if the 20 changes had taken place in different chromosomes, and if the variations were linked at all closely in inheritance the number required would be staggering. But there are breaks in linked inheritance, and the possibility is as stated. Associated with this benefit arising from the law of recombination, there is another of great practical importance resulting from the phenomenon of dominance. Recessive variations may arise, which in the particular factorial complex existing in the individuals at the time of origin, would cause the possessors to be eliminated by natural selection. These variations, however, may be carried an indefinite number of generations in the heterozygous condition, thus multiplying the chances that they finally be combined with other factors in complexes which as a whole are desirable. These inestimable advantages remain even though it should be shown later that the more fundamental and generalized characters of an organism are not distributed by Mendelian heredity. Loeb [129] suggests that the cytoplasm of the egg is roughly the potential embryo and that the chromosomes, distributed as required by the breeding facts of Mendelian heredity, are the machinery for impressing the finer details. There is very little to be said for this point of view, though it may have use as a working hypothesis. But granting its truth, it does not detract from the benefits gained by the origin of sex; the majority of variations are comparatively small, changes in detail, the very kind which are known to be Mendelian in their inheritance.

Yet sexual reproduction in itself does not assure these advantages, though they are based upon it. *There must be means for the mixture of germ plasms.* This opportunity was furnished originally by bisexuality. Afterwards hermaphroditism was tried; and, though manifestly an economic gain, it was, on the whole, unsuccessful except as functional bisexuality was restored by self-sterility, protandry, protogyny or mechanical devices which promoted cross-fertilization. The prime reason for the success of sexual reproduction, then, as Weismann [209] first maintained, though he knew not the exact reason, is the opportunity it gives for mingling germ plasms of different constitutions, and thereby furnishing selective agencies many times the raw material producible through asexual reproduction. It was not sexual reproduction *per se* which triumphed, but exogamy.

While increased variability and the greater elasticity in adaptiveness to new environments thus gained must be given the first consideration when seeking the significance of sex, they are not the only advantages. As we have seen, an increased size, greater viability and increased production of offspring commonly result from crossing somewhat different forms. Here is a combination of qualities unquestionably having survival value in the great majority of cases. It is a phenomenon so universal, so uniform in its effect, it must have played an important rôle during the course of evolution. Heterosis increasing growth and fertility immediately, segregation favoring adaptibility in the next generation, is a partnership of some strength. An income for life and a trust fund maturing for benefit of the children, what more could one ask?

Heterosis may even be pictured as the efficient cause of sex survival. Some means of favoring union of dissimilar spores occurred as a chance variation. Through the combination of somewhat different qualities this new dual product, the zygote, was better enabled to develop and to reproduce. Its survival coefficient was high. The tendency for union of spores persisted and became characteristic of the species. Sex was established.

This is a pleasing theoretical picture, and we do not believe it is overdrawn, but it must be admitted that the concrete evidence of a sexual union being immediately beneficial in the lower organisms is not what might be desired. Jennings [108] finds a marked slowing down of the reproductive rate in the generations immediately following conjugation in Paramecium, with no beneficial effect resembling heterosis, although he suggests that recombination may account for the *best* cultures. In fact, nothing similar to heterosis has been found in unicellular organisms. The lowest type where distinct evidence of the phenomenon has been discovered is in *Trochelminthes*, cross-fertilization increasing size, vigor, viability and reproductive rate in rotifers. But it would be strange indeed if no such effect did occur in low forms when it is so widespread in all the higher plants and animals. Heritable variations are constantly arising in simple organisms, as has been demonstrated by Jennings [107] in Difflugia, and it may be assumed that these are in part favorable and in part unfavorable. The union of two individuals would have the same chance of bringing together the greatest number of favorable growth factors and the progeny would thus be benefited, even though the mechanism for bringing this about is not as well organized as in the higher forms.

Some evidence of the possible importance of heterosis in the establishment of sex may be obtained by the consideration of an analogous phenomenon, double fertilization among the angiosperms. In the gymnosperms the embryo develops from the fertilized germ cell, of course, but the endosperm which nourishes the young seedling is gametophyte tissue. In the angiosperms the endosperm as well as the embryo develops after a fertilization has taken place. The conditions are slightly different, as a fusion between two maternal nuclei occurs before the union with the second male nucleus, but the essential feature is the same as in the production of the embryo—different hereditary materials are united when cross-fertilization occurs. And in the same way that the embryo and the resulting plant may be greatly benefited by cross-fertilization, so also is the endosperm tissue increased in amount as a direct manifestation of hybrid vigor.

Němec [162] has sought to account for endosperm hybridization as an adaptation which results in a better adjustment of the composition of the reserve food supply to the needs of a hybrid embryo. The cross between somewhat different types results in an embryo which presumably partakes of certain features of both parents. If forced to depend upon food supplied by only one parent it might be handicapped to some extent in comparison with another embryo supplied with food which was intermediate with respect to the two parents. If endosperm hybridization does indeed supply such a need, the fact that the endosperm is also increased in amount would have equal importance. It may well be that to fill either purpose endosperm hybridization has sufficient value to account for its maintenance in the angiosperms. However this may be, increased adaptability through recombination of characters which is such an important factor in sexual reproduction has no significance in this case, as the endosperm does not perpetuate itself.

Additional light may be thrown on the importance of heterosis in sex origin from the part it possibly has had in a related series of events. In the algæ and mosses, the principal life processes are carried on in the haploid generation and the parts which result from fertilization and produce the spores are relatively insignificant and are dependent upon the gametophyte for maintenance. How the sporophyte has gradually become more specialized, taking up the manufacture of food for itself until finally the relations are changed completely, are matters of common knowledge. This series of events is usually referred to as the rise of the sporophyte and decline of the gametophyte.

Just why there has been this radical and complete change in the plant kingdom is rather difficult to explain, but it should be noted that the increased variability and greater adaptability which seems reasonable in accounting in a large measure for the survival of sex, is not applicable here. Recombinations occurring at the reduction division can be utilized by the gametophyte as well as by the sporophyte, hence there seems to be no necessity for plants to change from dominant gametophytes to dominant sporophytes in order to secure the greater adaptability offered by sexual reproduction. Every combination of characters possible in the sporophyte occurs in the haploid condition, if we leave out of consideration heterozygous combinations which are the interaction of two members of a contrasted pair of factors and cannot be fixed. Haploid combinations even have certain advantages over diploid combinations in that all the characters are expressed and offer more material to natural selection. Favorable combinations have a better chance of survival and unfavorable combinations are more quickly eliminated. If, then, variability is not a factor in the rise of the sporophyte, and if we refuse to admit any value in chromosome-doubling itself, and the evidence certainly does not indicate that it has any significance, the only factor which remains, as far as we can see now, is the vigor derived from hybridization. Heterosis, of course, can operate only in the sporophyte. In the lower plants where the sporophyte is less important in the life cycle, heterosis would be of value only in spore formation. Later as the sporophyte became of more consequence, heterosis would have had more and more value; and it may well be that it had considerable to do with this revolution in plant life.

If sexual reproduction is so useful that it has been adopted as the principal means of reproduction at a sacrifice of speed of multiplication and economy of material, why, then, has it been given up by the many species which have resorted to vegetative propagation or parthenogenesis? Even self-fertilization, which is the rule with many plants, nullifies the advantages which were responsible for its development. As far as there is significance in amphimixis in inducing variability, continuous self-fertilization must for the most part be left out of consideration. Weismann[209] states the problem:

> If amphimixis has been abandoned in the course of phylogeny by isolated groups of organisms, this has happened because other advantages accrued to them in consequence, which gave them greater security in the struggle for existence; but it must be admitted that they thereby lost their perfect power of adaptation, and that they have thus bartered their future for the temporary securing of their existence.

Let us see what it is for which these organisms "barter their future." According to the view of heterosis outlined previously, there is no advantage in the heterozygous state in itself, but on account of linkage it is difficult to obtain all the more favorable characters which exist in a species combined in one individual in a pure breeding, homozygous condition. There is always the possibility of obtaining such combinations, however, and the resulting individuals are well fitted for survival as long as the environment remains the same. If the production of these favored few is accompanied by any change which renders cross-fertilization difficult, and if there is nothing to prevent them from resorting to self-fertilization, parthenogenesis or vegetative means of propagation, there is no obvious reason why the plants should not undergo the change. They would possess the most efficient means of multiplication and would doubtless be fitted for survival through long periods of time. They would not be flexible, however, and if the environment changed would probably lose in the race with more adaptable cross-fertilized forms. Their handicap is their lack of chances for progress.

A secondary advantage of sexual reproduction is the division of labor made possible by secondary sexual characters, using the term very generally and including even such differences as those which separate the egg and the sperm. It is not known just how these differences arose or by what mechanism they are transmitted. The greatest hope of reading the riddle lies in an investigation of hermaphroditic plants, for there are technical difficulties which till now have prevented its solution in animals. For example, breaks in the linkage between sex-linked characters occur only in the female in Drosophila, and as the sex chromosome is double in the female, it cannot be determined whether the differentiation between male and female is due to the whole chromosome or not. But this ignorance does not give reason for a denial of the great advantage which sexes bearing different characters hold over sexes alike in all characters except the primary sex organs.

The only glimpse of the truth we have on the matter comes from recent work on the effect of secretions of the sex organs on secondary sexual characters. The effect of removing the sex organs and the result of transplanting them to abnormal positions in the body have shown that in vertebrates the secretions of these organs themselves activate the production of the secondary sexual characters. This does not seem to be the case in arthropods, however; so one cannot say that primary sexual differentiation and secondary sexual differentiation are one and the same thing. Nevertheless, the generalization is not improbable. Surgical castrations of insects which have not affected the secondary sexual characters even though made in the early larval stages, are not conclusive because of the possibility of the primary sex organs having a marked influence on development very early in the life cycle; and parasitic castration brings in another variable through the presence of the alien organism.

Again, there is a presumable advantage in bisexual reproduction in having sex-linked characters. We say presumable advantage, for all of the relationships between sex and sex-linked characters are not clear. The facts are these: One sex is always heterozygous for the sex determiner and the factors linked with it. Even if there be no actual advantage in the heterozygous condition, if heterosis prove to be only an expression of the meeting of dominant characters, a possible advantage still accrues to this phenomenon because the mechanism contributes toward mixing of germ plasms. As an example, let us take the Drosophila type of sex determination. There the sperm is of two kinds; the one containing the sex chromosome and its sex-linked factors, the other lacking it. The eggs are all alike, each bearing the sex chromosome. It follows, then, that the male always receives this chromosome from his *mother* who may have received it from *either* her *father* or *mother*. Moreover, further variability may be derived from the linkage breaks which occur always in the female. This last phenomenon is hardly worthy of special mention, however, until it is shown to be typical of bisexual reproduction.

This short reconnaissance presents only the facts on the rôle of reproduction in evolution as they are affected directly or indirectly by inbreeding and outbreeding. A very great number of interesting things connected with reproduction during the course of evolution have not been mentioned. This is because it is felt that the vital feature in the whole affair, the persistence in both the animal and the plant kingdoms of innumerable mechanisms providing for cross-fertilizations, is to be explained solely on the ground of offering selective agencies the greatest amount of raw material. Mendelian recombination is thus assigned a part in phylogenetic development second only to inherent variability, and the whole history of reproductive change becomes clear without the ill-advised assumption that complex processes like autogamy are harmful in themselves.

CHAPTER XI

THE VALUE OF INBREEDING AND OUTBREEDING IN PLANT AND ANIMAL IMPROVEMENT

The origin of our more important domestic animals and cultivated plants is a matter on which there is no direct evidence. Among animals the ostrich is the only example of modern domestication; among plants not a single species of great economic worth has been brought into cultivation within historic times. If one must have a theory concerning their genesis, and what one of us does not delight in theorizing, the weight of evidence is in favor of a poly-phyletic origin in nearly every case. *There is more than one wild species related to our modern dogs, cattle, swine and sheep, our wheats, barleys, apples and grapes; and these species will cross together and yield partially fertile hybrids.* The wild relatives of the domestic forms were variable, so variable that many species were differentiated by natural causes; yet these species groups remained so well adapted to each other germinally that their hybrids are not completely sterile. What seems more reasonable than to suppose the original domestic races to have been produced by uniting two or more wild types and following this union of diverse germ plasms with more or less close inbreeding and selection?

Such procedure, at least, has been the method whereby the clearly distinct and highly valuable breeds of the present day have originated. Take the draft horses as an example. In the early days of Europe native breeds were developed in every country for military purposes. Just

PLANT AND ANIMAL IMPROVEMENT

how they originated we cannot say. The obvious fact is that none of them developed outstanding merits except the Flemish horse. Then improvement became rapid and steady. With an infusion of Flemish blood came the remarkable development of the Clydesdale in Scotland, the Shire in England, and the Belgian in the low countries. Adding the Arabian blood which came in with the defeat of the Saracens in 732, and the wonderful Percheron of France came into being.[75]

Similarly the origin of all modern breeds of coach, light harness and saddle horses may be traced. To the native breeds of Europe were added the blood of the Barb or its derivatives, the Turk and the Arabian. In France, in Spain, in England and in Russia the history is the same—hybridization, then close breeding and selection.

If one turns to cattle, the story varies but little. The basis of our modern strains is the cross between domesticated progeny of wild European cattle and their Asiatic relatives. From this stock numerous breeds grew up differing in contour, size and color. Some were horned, others were hornless. Some were developed for meat production, large at maturity and quick in attaining it; others were selected for the dairy, a great milk production and a high percentage of butter fat. As time went on and commercial channels became better established crosses were made between the better animals of the different beef breeds and between those of the various dairy breeds. Crossing followed by inbreeding has been the touchstone of success.

Similar more or less useless generalities could be given about swine, sheep, dogs, cats, the cereals, the perennial fruits, the numerous floricultural novelties, but this would serve no purpose. We have seen from our consideration of the facts of heredity that both inbreeding and outbreeding must be used if one would succeed in improving the products of domestication. There must be cross-breeding to furnish a variety of character combinations from which to select; there must be inbreeding to provide the opportunity to isolate the combinations desired. What we want to know now is the manner of their use, the degree of inbreeding permissible under given conditions, the efficacy of crossing for particular purposes.

While there has always been a certain amount of inbreeding as a necessary adjunct in building up breeds of livestock because of the necessity of mating near relatives in order to establish uniformity, the opinions of breeders have differed and still differ as to how long or how close intermating can be practiced with safety. Yet some of the most noted modern livestock strains owe their excellence to a close and continuous inbreeding that would be looked upon with misgivings by the majority of animal raisers. In fact, some of the inbreeding actually practiced was due more to enforced isolation, or the expense or difficulty of securing unrelated animals with desirable characteristics, than to a firm belief in the desirability of the method. This might be said of the Shetland pony, the Angora goat, the Merino sheep in America, and of many breeds of dogs.

Notwithstanding these facts, it would be a mistake not to recognize how great an amount of continuous and extended inbreeding has been practiced intentionally with the best of results after the general characteristics of a breed have been established. This is true as a generalized statement for the modern trotting horse and saddle horse which have shown so much speed; for the Shorthorn and Hereford, the most famous English breeds of beef cattle; for the Southdown and the other famous sheep breeds, the Shropshire, the Oxford and the Hampshire, to which it has given rise; and for almost all of the more famous breeds of dogs, not even excepting the large types, the mastiff, the St. Bernard, and the Newfoundland, which are derived from the Tibetan dog, *Canis niger,* as a foundation stock.

Perhaps the most notable examples of conscious use of intense inbreeding in developing breeds of marked excellence are the dairy cattle of the channel islands, the Jersey and the Guernsey. One does not need to describe or to eulogize these strains. What they are and what they have accomplished in producing milk and butter fat are known throughout the world. Starting with the cattle of Normandy and Brittany as foundation stock, these two breeds have been built up by persistent use of a more intense system of inbreeding than is recorded in the history of any other strain of livestock. In fact, since 1763 rigidly enforced laws have prevented landing any live cattle whatsoever on either island except for slaughter. When one realizes that the larger of these two islands, that of Jersey, is but eleven miles long by six miles wide, he can appreciate the amount of inbreeding these laws have promoted.

With swine, one gathers that injurious results from close mating may be somewhat more pronounced than with some other animals; in other words, that swine carry a large number of deleterious recessive characters. But many of the famous breeds of swine have been rather closely inbred. Mr. N. H. Gentry of Sedalia, Missouri, who has achieved quite a remarkable success with Berkshires, rarely went outside of his own drove for breeding stock. He is quoted as saying (Mumford [158]): "If it is true that inbreeding intensifies weakness of constitution, lack of vigor, or too great fineness of bone, as we all believe, is it not as reasonable and as certain that you can intensify strength of constitution, heavy bones, or vigor, if you have these traits well developed in the blood of the animals you are inbreeding? I think I have continued to improve my herd, being now able to produce a larger percentage of really superior animals than at any time in the past."

This quotation exemplifies the opinion of the best informed of the practical breeders of the present day in regard to the practice of inbreeding. In general they recognize that the results obtained depend largely upon the character and constitution of the animals, and the care and skill with which they are selected for mating. They have learned by experience what matings are the most successful and how far it is advisable to carry close breeding with a particular stock. Rarely is inbreeding as close as brother and sister or parent and offspring mating continued for many successive generations, however; for they are apprehensive at all times that inbreeding *may* reduce the fertility and lessen the constitutional vigor of their animals, and they frequently introduce stock from outside to counteract any tendency in this direction whether fancied or real.

In plants the problem is different. No systematic individual mating system is practiced, as is the case with animals, so that whether plants are inbred or outbred is a matter which is left to regulate itself automatically.

Among those plants which are largely self-pollinated by nature, chance crossing, or, in some cases, systematic hybridization, has originated new types. Self-pollination has brought these types to uniformity, and by isolation new varieties have been established. Among naturally crossed plants genetical variations are continually being produced and selection for certain of the more conspicuous features has led to the creation of well-marked varieties. Indian corn is one of the best examples in this class. There are many distinct types, and the less distinct but fairly well recognized varieties are almost innumerable, adapting the plant to a range of conditions from the edge of the Arctics to the Tropics, throughout the world.

In every locality where corn is grown the usual habit is to prevent inbreeding as much as possible. Many corn growers make a regular practice of bringing in seed from other localities, and often two or more somewhat different varieties are planted together and allowed to mix. The reason why this practice is followed is easily apparent from the controlled experiments on the effects of inbreeding and cross-breeding upon this plant. But even keeping in mind the injurious results of inbreeding, indiscriminate crossing is not desirable. Many of the well-known varieties in the Corn Belt States, such as Reid's Yellow Dent, Leaming, and Boone County White, are the results of long-continued selection for certain standards without crossing with other varieties. Inbreeding, therefore, has secured individuality for varieties of cultivated plants as well as for breeds of animals.

The value of inbreeding in plant and animal improvement in the past may be summed up in the statement that it is the greatest single agency in bringing about uniformity and the concentration of desired qualities. So valuable have been the results, particularly with animals, that it has often been continued even though concentration of characters which made for lessened constitutional vigor and fertility accompanied the accumulation of desirable features, for the good outweighed the evil. To overcome anticipated calamities, animal breeders have from time to time introduced fresh stock. In doing this they certainly were wise, since a rather high probability always exists that such a procedure will introduce the dominant complements of the detrimental characters. But even granting the good sense at the base of both practices, it may be doubted whether inbreeding and crossbreeding have been used in the best possible manner as means of improvement. There are precise uses to which each may be put which hitherto have not been considered.

Experiments with maize show that undesirable qualities are brought to light by self-fertilization which either eliminate themselves or can be rejected by selection. The final result is a number of distinct types which are constant and uniform and able to persist indefinitely. They have gone through a process of purification such that only those individuals which possess much of the best that was in the variety at the beginning can survive. Although these resultant, purified types have little value in themselves, they have possibilities. The characters which they have can now be estimated more nearly at their true worth. By crossing, the best qualities which have been distributed to the several inbred strains can be gathered together again and a new variety re-created. Size, vigor and fertility can be fully restored with the advantage of real improvement through the elimination of certain undesirable characters.

At present, this application of inbreeding to the improvement of cross-bred animals and plants is somewhat of an unknown quantity. It has not been as thoroughly tested as might be desired, but the basic principle is sound. Although it is a drastic procedure, it is merely utilizing to the fullest extent what practical breeders have recognized as one of the most valuable benefits of close mating. Accepting the doctrine that consanguinity in itself is not in any way injurious and that good or evil results from it solely through the inheritance received, we can attack the century-old problem of inbreeding with a clarity of vision heretofore impossible. Breeds of animals, and naturally crossed varieties of plants, which are necessarily more or less heterogeneous in their hereditary constitution, can be split up into their component parts by this means. The pure types obtained can then be selected with far more surety than is ever possible with organisms in a continuously hybrid condition, thereby presenting basic stock of tested value for further hybridization and recombination.

With plants the application of this method would be simpler than with animals. Most naturally crossed plants can be artificially self-fertilized and constancy and uniformity reached in about eight generations if there are no complicating factors such as self-sterility. The expense would not be prohibitive, although many pure lines must be tested in order to have a high probability of obtaining all that is best in a variety. After the most desirable combinations are isolated, their recombination into a new and better variety, which could be maintained by seed propagation, would be a comparatively easy undertaking.

With plants which are propagated vegetatively, the matter is even less difficult. Nearly all varieties of fruits, flowers and vegetables propagated in this way are notoriously unstable when grown from seed. The excellent varieties that we now have undoubtedly owe their superiority in large measure to a fortunate combination of many different characters so made as to obtain the maximum effect from hybrid vigor. Attempting to obtain further improvement by crossing these already widely crossed varieties is like trying to solve a picture puzzle in the dark. First analyze the material to be used by systematic and rigorous inbreeding, let the consequences be what they may. Then cross the different constant types which may be ultimately obtained and test one combination after another until a real improvement is effected. When that is done the individuals can be propagated indefinitely by the same means utilized before. Of course, this method has the objection that many of the plants propagated asexually require several years for each sexual generation. Results would be slow for that reason, it is true, but they would be sure.

With animals the application of this method would be quite a different proposition. Inbreeding closer than brother and sister mating could not be practiced, and the time required to obtain purity and constancy would be much greater than is the case with self-fertilization. Moreover, the number of individuals which could be obtained would be so small that selection could not be made advantageously. Finally, the cost of raising most animals is so great that the maintenance of animals of little or no value in themselves solely for a possible ultimate improvement might well be too discouraging an undertaking. But what could be done is to use animals from some of the intensively inbred herds of the present day as basic stock for building up new strains through crossbreeding and selection. The point which we particularly wish to make here is that the apparently disastrous effects of inbreeding need not be so greatly feared as is usually the case; because if anything is lost by inbreeding it is usually something undesirable. Inbreeding, therefore, may prove to be a very great gain if used as a method of purifying and analyzing a cross-bred stock.

While the full value of inbreeding in plant and animal improvement has not as yet been fully recognized, the advantages derived from outbreeding are more generally known. Outbreeding as a means of improvement may be considered under two heads: First, the immediate value to be derived from crossing related types and thus securing the maximum benefit from hybrid vigor; second, the more complex problem of crossing radically different forms to create variability out of which new breeds or new varieties may be constructed by a process of selection.

In some cases the first generation cross, although vigorous, is sterile. An example is the mule, which, though having the disadvantage of not being able to reproduce, has held a place in agriculture and industry throughout historic times. According to Mumford [156] there were nearly five millions of these animals in the United States in 1915. Of it he says:

This was more than one-fifth of the total number of horses in the country at the time. The production of mules has increased at a more rapid rate than horses, and the use of mules is becoming more extensive. The mule hybrid is a remarkable example of the practical advantages which follow a particular cross. This animal is more hardy and enduring than either parent. As compared with the horse, the mule is longer lived, less subject to disease or injury, and more efficient in the use of food. The mule can be safely put to work at a younger age, will thrive on coarser feed, and seems to be much better able to avoid many dangers which menace the usefulness of the horse. The mule will perform more arduous labor on less food. The mule will endure the heat of southern latitudes more successfully than the horse and is therefore a more popular draft animal in the South.

Other first generation crosses among animals, which are not sterile like the mule, have good qualities and are well known. Youatt, early in the nineteenth century, stated that crosses between the English and Chinese breeds of swine were frequently made, and that in Germany the native breeds were often crossed with the English breeds. To-day the first generation cross between the Duroc-Jersey and the Poland-China, and between the Poland-China and Chester White are popular animals among the feeders. No attempt is made to breed from them as it is well known that the later generations are variable in color, size and conformation, and usually possess less vigor than the animals of the original cross.

First generation crosses between many of the standard breeds of beef cattle are raised, and frequently they win the first prizes at the stock shows. The Shorthorn and Aberdeen-Angus combination is popular.

The Mediterranean breeds of poultry are sometimes crossed with the heavier types. First crosses of Leghorns and Plymouth Rocks give birds which are not so apt to become over-fat and yet are more valuable for meat than the smaller Leghorns.

The opportunities for improvement in this way

through the utilization of hybrid vigor are no less great in plants. The increased cost of seed is an item and the practice can only be followed with those plants which are easily crossed and which produce a large amount of seed. Many plants in which production might be increased in this way have such low economic value, however, that it would not be profitable to utilize the method. Cases in point are squashes and pumpkins. Tomatoes and cucumbers in certain crosses, on the other hand, have been found to give appreciable increases in yield and other desirable qualities, advantages which are readily secured every time the particular cross is made.

Maize is the plant which is most suitable for use in this way, a notable fact since it is the most valuable farm crop in the Western Hemisphere. The reason it merits this statement is because it is easily crossed on a large scale by sowing the two types to be crossed in alternate rows in an isolated plot and detasseling all of one kind before pollen is shed. As early as 1876 Beal[3] reported that corn could be increased in yield in this way. Since that time numerous tests have been made and the fact is established that crosses between varieties of corn of somewhat different type may be expected to outyield either parent in many cases, and when the parental varieties differ in time of maturing may be expected to ripen earlier than the later parent. Thus out of fifty first generation crosses between varieties of corn grown in Connecticut, eighty-eight per cent. yielded more than the average, and sixty-six per cent. yielded more than either parent. The average increase in all the crosses above the average of their parents was about ten per cent., including the crosses which gave no indication of hybrid vigor.

The greatest increases occurred in the crosses between flint and dent varieties, and often there was a really noteworthy hastening of the time of ripening, which is of considerable importance in those regions where early fall frosts are a limiting factor.

The greatest improvement to be made in this way comes from crossing varieties which have previously been put through a process of self-pollination. When certain inbred strains are crossed the increase in growth is remarkable, as previously noted. This comes partly from the fact that following inbreeding the maximum effect of hybrid vigor is obtained while in ordinary varieties segregation brings about partial homozygosity in many plants. It is also due to the elimination of many undesirable characters during the process of inbreeding. The crossed plants are remarkably uniform. One plant is a replica of another. Given proper conditions they all produce good ears which form a remarkable contrast to ordinary varieties in their similarity to each other. There are no barren stalks, and the abnormalities and monstrosities which commonly occur in every field of corn are almost entirely absent. In those cases in which one or both of the parent strains is resistant to parasitic infection, such as smut, the cross is also resistant and this is a factor for greater production.

There are, on the other hand, certain serious disadvantages in the practical utilization of first generation crosses between inbred strains. In the first place the yields of the inbred plants are low, which makes the cost of the crossed seed high. What is more serious, the seeds produced on inbred plants are small and less well developed than seeds of ordinary corn, and the seedlings coming from these seeds are less vigorous and are thereby greatly handicapped at the start. The plants at first are smaller and have a less healthy color than plants of ordinary varieties, and although they usually overcome this handicap, they may not always do so if the conditions in the earlier part of the season are particularly unfavorable.

A method which overcomes these objections is now being tested at the Connecticut Agricultural Experiment Station, and promises excellent results. This method is as follows: Four inbred strains are selected which when tested by crossing in all the six different combinations give an increased yield. Two of these strains are crossed to make one first generation hybrid and the other two are crossed to give another. These two different crosses, which are large vigorous plants, are again crossed and the seed obtained used for general field planting. This procedure may be diagrammed as follows:

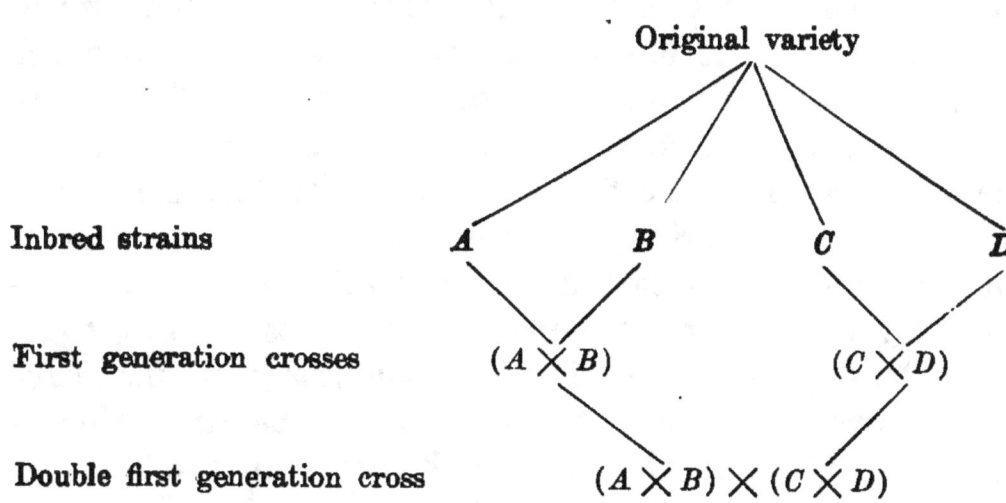

In this way large yields of well-developed seed are obtained, and the young plants are not handicapped in any way. The beautiful uniformity of the first cross is sacriticed, but the advantages gained promise to counterbalance any loss in this respect. Theoretically there is little reduction in heterozygosity and presumably little reduction in the incentive towards increased size and productiveness. A great many different possibilities are involved in such double crossing and they have not been sufficiently tested to warrant extravagant claims, but judging by their appearance such doubly-crossed plants are clearly the finest specimens of corn so far obtained under the conditions in which they have been tested.

The first impression probably gained from the outline of this method of crossing corn is that it is a rather complex proposition. It is somewhat involved, but it is more simple than it seems at first sight. It is not a method that will interest most farmers, but it is something that may easily be taken up by seedsmen; in fact, it is the first time in agricultural history that a seedsman is enabled to gain the full benefit from a desirable origination of his own or something that he has purchased. The man who originates devices to open our boxes of shoe polish or to autograph our camera negatives, is able to patent his product and gain the full reward for his inventiveness. The man who originates a new plant which may be of incalculable benefit to the whole country gets nothing— not even fame—for his pains, as the plants can be propagated by anyone. There is correspondingly less incentive for the production of improved types. The utilization of first generation hybrids enables the originator to keep the parental types and give out only the crossed seeds, which are less valuable for continued propagation.

The second phase of the subject of outbreeding in its relation to plant and animal improvement—that of wide

PLANT AND ANIMAL IMPROVEMENT

crossing between distinct varieties, species or even genera —is so large a topic it cannot be more than touched upon here. Each particular cross presents technical problems of its own. All one can say as a generality is that the principle in every case is the same. Crossing brings together germ plasms having various attributes. These attributes, the hereditary factors, recombine with regularity and precision. They Mendelize. From Mendelian segregation and recombination come the possibilities of new and improved races. Except in those rare instances when new variations previously unknown to the species occur, nothing can come out of the cross that did not go in. But the number of combinations possible when the two parents differ by many hereditary factors is so great that practically speaking many character complexes may appear which have never before had the chance of showing their merits or defects. In them lie our hopes.

It was noted earlier that many species crosses are partially sterile, that there is often a degeneration of many of the germ cells and embryos, and that certain extreme types are thereby produced more frequently than is usually to be expected. The extreme variability induced by such wide crossing offers the best field in which to look for the beginnings of new and valuable types of animals and plants. This is not a theory; it is a general fact born of long experience, for when we look into the origin of many of our most valuable domesticated animals and plants we find unmistakable evidence of their hybrid ancestry.

CHAPTER XII

INBREEDING AND OUTBREEDING IN MAN: THEIR EFFECT ON THE INDIVIDUAL

THE world has entered an age of reason. The leaven of education is working rapidly, and all relations of man to his fellow-man, all connections of man with his environment, are being subjected to thorough scrutiny. To accompany the current changes in the arts due to new advances in science, enlightened democracy demands progress in religion and philosophy, in government and social policy. It has set upon its program the task of establishing a broad scheme of social hygiene, and more than one might suspect has been accomplished.

Although there is still room for improvement, general communistic sanitation has reached a degree of efficiency which a few years ago would hardly have been deemed possible. The civilized world has gone through a cleaning-up period which has provided reasonably hygienic buildings, tidy streets and excellent waste disposal; which has bettered the condition of the people and lowered their death rate by quarantine regulations, public hospitals, and free medical attention in the schools; and has passed on to preventive work, vaccination, pure food legislation, and the like. There has been marked progress in ameliorating conditions of work. Sanitation has been made the subject of many laws; hours of toil shortened—particularly for women and children. Factory supervision and wage regulation are accepted facts; industrial insurance is in the air. Public education has made strides with seven league boots. That dignified monument, the free school, is not the only evidence. There are normal schools and universities, museums and research institutions, public collections of books and public printing of books in numbers sufficient to form libraries by themselves.

Is it realized just what this means—why social policy has developed in precisely this manner? It is because this is the mental line of least resistance, the order of social reform needing the least foresight. The first efforts were to clear up obvious filth, the accumulated débris of human activity—the record of the past; the step forward was an appreciation of the efficiency in production resulting from comfort and satisfaction in conditions of work—the present; and then came the spread of educational facilities—a preparation for work, an insurance on the immediate future.

But change, progress, reform, whatever one may call it, ought not and will not stop here. The program of social hygiene is not complete if there is failure to provide for a future still more distant. And this is the real thought in the minds of a few clear thinkers of Europe and America whose names are connected with the spread of eugenic policies. It was thought for the care of the coming generation that led Budin to establish the *Infant Consultations* and *Milk Depots* in Paris, that led Miele to start his School for Mothers in Ghent. It was thought for the future of the race as a whole that gave the impulse to Galton's work.

We have no eugenic system of conduct to lay down here, for we believe the acquisition and diffusion of knowledge are needed more than widespread dogma and ill-advised legislation at the present day. The recommendations of the sympathetic altruist with a little learning have done more than anything else to hinder a healthy growth of eugenic ideas. All we would ask is that the physician, the clergyman, the social worker, the penologist, the statesman, give conscientious consideration to the facts of heredity as a guiding principle in the solution of the problems of the family with which they have to do. No questions are so hedged about with superstition, with irrational tradition, with religious dogma, as those which concern sex and reproduction; no problems are more delicate, more difficult, than those which seek the direction of human evolution; yet after all man is an animal and must be dealt with as such. Civic law he may escape, to natural law there is no immunity.

We have seen how characters are transmitted in sexual reproduction in the lower animals and in plants, how hereditary differences carried as potentialities in the germ cells are shuffled and divided when these are formed, by a law as definite and precise as one of chemistry or physics. We have seen how the operation of this law brings about the outstanding phenomena of inbreeding and outbreeding. Man is just another sexually reproducing mammal and *a priori* his heredity is guided by this law. Being a thoroughgoing egotist, he doesn't like to realize this. It takes time for the truth to filter in. Comparative anatomy and physiology and the doctrine of evolution have been the greatest agents in this familiarization process. The veriest schoolboy now recognizes the homologies between the bones and muscles of the lower mammals and those of man, and sees nothing out of the ordinary that their digestive metabolisms are the same.

Those with no biological training have now no difficulty in accepting as fact the idea that man came into being by the same process of evolution as the rest of the organic world. But even in these cases it has been a long struggle against prejudice, and the scientific study of heredity is too recent to have outgrown it. We will, therefore, not confine our argument strictly to the logic of the question. Inheritance in man has actually been studied by the same general methods as have brought such wonderful results in other organisms, and corroboration of every detail has been the outcome.

When one says the fundamentals of Mendelism have been supported in detail by investigations on the human race, he does not mean to imply that the critical investigations needed to establish the Mendelian hypothesis in the beginning were supplied by such data. This is obviously impossible because of inability to control matings. The only records which can be analyzed are pedigrees of families carrying some striking hereditary phenomenon. Such a method is unsatisfactory because the data must be gathered second-hand through several generations—often by untrained workers. It is necessary to work backward instead of forward, to be content with fragmentary information, to realize the high percentage of experimental error. What is meant by corroboration of Mendelism in human heredity is simply that starting with the assumption of the truth of the law, all human data have been found to fit. But it is often very difficult to say whether the inheritance of a particular human trait is dominant or recessive, whether it is controlled by one or by several factors, whether it is sex-linked or independent.

Several skeletal abnormalities unquestionably show a

high degree of dominance. Among them may be mentioned the peculiar type of dwarfing known as achondroplasty, and the various digital malformations termed medically brachydactyly, polydactyly and syndactyly. Evidence of complete dominance is probably better in these than in any other cases, but in view of the many instances where subsidiary factors enhance or diminish the expression of a primary factor, it seems decidedly unwise to follow Davenport and to recommend marriage with unaffected members of such families with the assurance that the latter cannot transmit the trouble which afflicts their relatives. If this advice could be accepted in good faith, the inheritance of dominant traits, whether disagreeable or desirable, would have little interest. They would stand revealed in those possessing them; they alone could transmit them. But this is not the whole truth. Some of the so-called dominant characters in man are abnormalities which no one cares to see expressed in his or her children, and their dominance is imperfect or uncertain. In many instances the records have been analyzed hastily and carelessly; for example, hare-lip and cleft palate, which is clearly a recessive condition in face of the data, though passing as dominant in the various textbooks of heredity. In all cases there is no guarantee that the unaffected member of the stricken family is germinally a pure normal. The family is one whose alliance is not to be sought by those who have a proper pride in a normal healthy posterity. Let us enumerate some of the troubles that come in this category: Hereditary cataract, ichythyosis or scaly skin, defective hair and teeth, diabetes insipidus, Huntington's chorea—an affection of the nervous system, imperfectly developed sex organs.

Does any one desire the establishment of sub-races thus characterized?

Other undesirable traits are more certainly recessive and the heterozygous carriers of the factors which control them cannot be distinguished by any differentiating characters of their own. Some of these abnormalities are extremely rare and for various reasons are not likely to increase. Among them may be mentioned pigmentary degeneration of the retina, Friedrich's ataxia, and xeroderma pigmentosum. But there are others which well may give some cause for dismal forebodings—hereditary feeble-mindedness and some forms of epilepsy and insanity. These characters may be put down as largely hereditary, and probably transmitted as single Mendelian units, but it must not be supposed that each manifestation of them is of similar kind. From the graduated character of feeble-mindedness and from the frequency with which epilepsy and other forms of neurosis appear in feeble-minded families, it is reasonable to suppose that minor factors of several types play a part. Nevertheless, for the deductions we wish to make here, they may be accepted as true examples of Mendelian recessiveness.

Other characters are not so simple in their inheritance. The Davenports [42, 43, 44, 45] have collected a large amount of data on the inheritance of skin color in negro-white crosses, the inheritance of hair color in Caucasian mixtures, and the inheritance of normal differences in stature. These characters are all complex. They are transmitted just as are the differences in height in plants—more or less of a blend in the first hybrid generation, and the appearance of such second generation types as would be expected if the differences were controlled by the segregation and recombination of several factor pairs. This in general is the interpretation given the inheritance of general mental ability or inherent ability in music, literature, art, or mathematics. We simply know that such abilities are inherited in some complex way, which, it is logical to assume, is Mendelian. We know the *fact* from pedigrees of families in which ability of a particular kind is very marked; we make the *assumption* from such circumstantial evidence of the generality of Mendelian phenomena as has been presented in abstract in this volume.

Having this basis, what shall be said of the effect of inbreeding and crossing on the individual? It would be easy to point to the conclusions reached when discussing domestic animals and plants, and say: "The same line of reasoning holds for man; draw your own conclusions." But this is hardly satisfactory. It is true enough, as a generality, to point to the desirability of some mating outside a particular line in order to assure physical vigor by complementary hereditary factors meeting each other, or to mention the possibility of undesirable characters being brought to light in some strains and of desirable characters being added in others by inbreeding. One would hardly feel this to be an answer to the question. If the study of heredity has resulted in an advance in knowledge having some practical value, it ought to be possible to make a more definite analysis of the facts as applied to the human race.

Let us ask first, What is ability in the human race, and what the evidence that it is inherited? A fair definition of ability may be given in the phrase, "skill in accomplishment," and this puts considerable emphasis on mentality. We all desire a healthy mind in a healthy body, but a feeble-minded Goliath is hardly of much use in the world, while a Robert Louis Stevenson struggling through life with the handicap of a delicate constitution leaves an imperishable monument. At any rate, there are few who deny the inheritance of physical differences. Pedigrees showing the exact method of inheritance of physical traits are too numerous.

The first real study of the inheritance of mental capacity was Galton's "Hereditary Genius," published in 1869.[73] By comparing the attainments of the relatives of eminent men from the United Kingdom with the attainments of its population as a whole he proved beyond a reasonable doubt the inheritance of potential capacity, though he had no inkling of how this capacity was transmitted. His conclusions have been corroborated by the works of Havelock Ellis[60] on "British Men of Genius"; of Woods[223] on "Heredity in Royalty," where lack of opportunity did not play such a disturbing rôle; and of Cattell,[22] Nearing,[160, 161] and Davenport[40] on eminent Americans. Perhaps the most striking feature of Galton's researches is the evidence of rarity of genius among a people who have contributed the greatest amount of creative work of the first magnitude in modern times (see Merz[140]). Only two hundred and fifty men per million of the British population became eminent. Though unquestionably this proportion must be increased several times because of the lack of opportunity of those similarly endowed to give full rein to their capacity, and because eminence as measured by history is fallacious in the extreme, nevertheless, when translated into the terms of modern genetics this ratio has a definite meaning. The hereditary factors which contribute toward the possibility of genius are numerous. Only occasionally is the proper combination brought together. The factors exist in the population at large, distributed in part to one individual, in part to another, but in the main the combinations make but for mediocrity. Only on a rare occasion is a favored one so showered with these gifts that he stands out supreme among his fellow-men.

No one knows what the component parts of these desirable qualities are, or can distinguish by external traits the individual who carries them, but a knowledge of the operation of Mendelian heredity enables one to say in a general way what ought to occur under given conditions with the same confidence as when dealing with similar indefinite qualities in the lower animals. Close selection, inbreeding, tends toward the production of gametic purity with mathematical precision. Does any one doubt but that close breeding in families which have shown superior civic value tends to concentrate, to purify, in genetic terms to render homozygous, the particular factorial combinations which cause this superior endowment? Will any one deny there is a real privilege in being allowed to marry into a family of proved worth, or a real reason for that family to scrutinize carefully the ancestry of one who asks to become allied with it?

We have seen how when certain hereditary factors have been brought together by the proper breaks in linkage, they tend to be transmitted together in the same manner as did the previous set of coupled factors. This same idea, followed to its logical conclusion, is a great help in visualizing the inheritance of capacity. The individual who owes his capacity to a complex which may be represented graphically by the linked series *AbCd-aBcD* where the capital letters represent the desirable factors, must be much more common than the individual who by linkage breaks receives the inheritance *ABCD-abcd;* yet the latter is the one who has the greatest power of transmitting his endowments. Of the influence of the individual in heredity, much has been written, particularly by those who, great themselves, have founded great families in American history. Elizabeth Tuttle through Jonathan Edwards, William Fitzhugh, the three Lees—sons of Richard Lee, and the various lines established by John Preston—Venable, Payne, Wooley and Breckinridge—are examples. It seems most reasonable to suppose that such prepotency for good comes about by the gathering together of groups of significant factors in the manner outlined above. Can one object to the concentration of such worth by relatively strict inbreeding despite its possibilities for ill? In fact, if we examine carefully the geneological records of such families, marriage of near relatives *is* found to be a common occurrence. Would it not be wise to do away with statutes against the marriage of first cousins such as are laid down in the laws of nearly half our States, even though the argument on the other side, as we shall show, is just as great? If such laws had been followed in every mating the world would have lost an Abraham Lincoln and have been compelled to punish a Charles Darwin.

The mention of the name of the great Civil War President doubtless brings the question: How does one account for his capacity on this hypothesis? The question is well placed. Such geniuses as Lincoln, Dalton, Faraday, Franklin, Pasteur—scions of the common people, simple examples of greatness standing among the commonplace—

have been the stock arguments of those sociologists who believe every man a star of the first magnitude darkened by lack of opportunity. Let us consider such cases in some detail.

The worst trouble with the euthenic idealists is that superficially they are not so far wrong, although fundamentally they miss the point entirely. Greatness in this world is governed by many factors. Appreciating Lincoln and Franklin for all they were, it is still allowable to question whether there were not some thousands of others of the same periods who would have entered the Hall of Fame had they been given the same political environments. Many of these contemporaries doubtless were great in the callings to which they were turned by force of circumstances; but the world seldom makes a very wide path to the door of him who makes a better mouse-trap than his neighbor. Castle [18] has called attention to the notable astronomers, the eminent biologists, inspired by the teachings of Brünnow and of Agassiz. If Brünnow had found a home at Princeton instead of Michigan, and Agassiz a place at Yale instead of Harvard, new names would be found among American astronomers and biologists, but their number would not be less. In short, one may admit with the euthenist the rôle of chance, of opportunity, of personal influence, of political preferment, of economic stress, in the moulding of men; one may acknowledge the difficulties attending the ranking of the great and the near-great; and yet abate not a jot or tittle of the position that inherent capacity, inherited potentiality, lies at the base of all. It is the one solid foundation on which to build.

Could one gauge the ability of the progenitors of

Franklin and Pasteur in some other way than by the biographical dictionary, they would probably be found to have a fair share of natural gifts. Their family lines are not to be compared with those of the notorious Zeros, Jukes and Nams, where the individuals through their bad heredity simply lacked even moderate capacity and were unable to rise when given normal opportunity. They belonged to the good solid *bourgeoisie* stock which forms the balance-wheel of modern democracy. But even granting to historical rank a justice which it does not have, admitting no personal superiority in any of the relatives of these two men and others like them, their very existence is a link in the proof of Mendelian combination in the making of mentality. The great bulk of the population inherits certain factors contributing toward ability. They do not have any one of the numerous inherited complexes which make a genius in the rough, but they have a random sample of the constituent parts. As a mere fulfillment of the laws of chance, matings between these individuals must occasionally bring together the happy combination of which we speak.

Whether these talents lie wrapped in a napkin or not, depends, of course, on a variety of circumstances. One *can* keep a good man down, though with some difficulty. It was probably the general attitude of society in those particular epochs that gave a Golden Age to Greece, a Renaissance to Italy, an Elizabethan period to England, a Napoleonic era to France, rather than a concentrated production of high mentality. Galton concluded the ablest race in history was that built up in Attica between 530 and 430 B.C., when from 45,000 free-born males surviving the age of fifty there came fourteen of the most illustrious men of all time. He was hardly justified in this. Mentality was then the order of the day in Attica. Accomplishment was appreciated. Minds of high order were drawn from surrounding countries. The great poet, strange as it may seem, was valued more highly than the wealthy merchant. Such conditions must account for much of this apparent racial superiority. Further, there can be little question but that in this little settlement there was much selective breeding. Had we the data, would we not find the Athenians all more or less related to one another? Had they not built up somewhat of a super-stock by inbreeding? Endogamy was their custom (Westermarck, 1901). Marriage with half-sisters was allowable, and if an Athenian lived as husband or wife with an alien, he or she was liable to be sold as a slave and have all property confiscated. Such inbreeding, given the possession of desirable characteristics on which to base selection, could hardly fail to bring results.

In a sense this is the obverse of the picture; the reverse is not as pleasing. Dreary histories have been written of consistently degenerate families, families with such a monotonously infamous record they are known throughout the world. There are the Jukes,[48] an inbred family whose record of pauperism, prostitution and crime has been traced for six generations. There is the "Tribe of Ishmael," a race of indigent vagrants since 1790, consistent in their ways of life no matter what their surroundings.[132] There is the Nam family, descendants of a racial mixture, Indian-white, less uniform than the first two in their anti-social traits, but characterized on the whole by vagabondage, stupidity and lack of ambition.[65] There is the family of which Poellman recorded 709 life-histories, a distressing chronicle of illegitimacy and pauperism. There is the Zero clan, a name fixing well their value to the world, ne'er-do-wells since the seventeenth century.[115]

We have no intention of going into further detailed descriptions of these people. They are but examples of an heredity which the world does not desire. He who does not believe their characteristics to be due to their germinal constitution allows his sympathies to run riot with his reason. Let anyone examine the published pedigrees of these tribes of unregenerates, in the light of the facts discussed in these pages. He will find degenerate mating with degenerate generation after generation, incest of the highest degree, inbreeding of the most intense kind. He will see segregation in different strains characterized by distinct kinds of degeneracy, as in the Jukes, where the descendants of Ada are prevailingly criminal, the offspring of Bell without sexual inhibitions, the progeny of Effie paupers. He will find that outbred lines have sometimes separated from the main stock, have had ambition, and have gone out and made respectable names for themselves. He will find that these people have been given chances, have been removed from old associations, taken to reputable homes, clothed, fed, educated according to their capabilities, and have remained—degenerate.

Social workers have been troubled because more representatives of these blood lines have not been removed from their isolation and the evil example it entails, and been given the stimulation of association with people of more stamina. Let us be glad that these natural experiments were carried on as they were. The disadvantage of neglect, isolation and bad example to the individual must be admitted, but does anyone believe that these families would have been a credit to the communities harboring them if the environment were changed. It was tried many times and failed. No! What happened in these cases was the establishment of near-homozygous races having a bad heredity. The result of inbreeding where the germ plasm is bad stands forth as a terrible example. What would have happened had there been no isolation would have been the contamination of good blood lines. In fact, certain illegitimate sub-strains in these clans did stand out above their relatives. Was this not due to a better endowment being brought in from the alien male?

The traits just discussed, at best mere uselessness or lack of capacity, seem to be somewhat less complex in their heredity than those leading to marked superiority, if one may judge by the seeming ease with which they are concentrated. Other undesirable mental characteristics appear to be still less complex. These are feeble-mindedness and the conditions related to it. Goddard[81] has studied 327 family histories in which feeble-mindedness entered. Somewhere between 50 per cent. and 75 per cent. of these family trees show distinct evidence of the hereditary nature of the defect. There were 144 matings of feeble-minded with feeble-minded producing 482 children of which all but 6 were feeble-minded. These few exceptions to the expectation for the union of two Mendelian recessives may reasonably be explained by assuming there is a paternity other than that assigned. Other types of mating come just as close to Mendelian expectancy, although Goddard himself has failed to analyze them properly by not correcting for the necessary omission of heterozygous matings having *no* feeble-minded children. We may, therefore, conclude that feeble-mindedness is due to a single principal unit factor, recessive to what we may call normal mentality.

There is evidence, however, of other minor factors which modify the grade of feeble-mindedness, and considerable reason for feeling that in some similar way certain forms of insanity, epilepsy and other neuroses are in some way related. At any rate, all of these abnormalities are in many cases inherited as recessive traits.

Here, then, is something wholly undesirable which may be the result whenever the proper unions occur; and as we have seen how inbreeding tends to bring out recessive characters, in feeeble-mindedness lies a potential danger. Let us see what this danger is as regards the United States.

It appears that in our present population of over 100,-000,000 there are something like 300,000 persons who are feeble-minded, epileptic or insane through an hereditary defect, a ratio of 3 per 1000. How many of these defectives resulted from a mating wherein at least one of the parents was of the same type is a difficult question and can only be answered with a rough approximation. The statistics at present available are meagre, but from their examination 200,000 may be considered to be above the mark. This leaves 100,000 defectives, then, which have been produced in a single generation by the mating of two transmitters of defective mentality who did not exhibit such defects in themselves.[56]

These 100,000 defectives were produced during a period when there were rather less than 20,000,000 married couples of reproductive age, by parents who were both heterozygous. But since only one-fourth of the progeny of such matings will be defective, at least 100,000 couples of this type were reproducing throughout this time. This low estimate would presuppose the survival of four children per couple long enough to have their mental status determined, an assumption which would require a total reproductivity of six or seven children per married pair.

If, then, out of 20,000,000 pairs of married persons, 100,000 were heterozygous for feeble-mindedness or attendant ills on both sides of the house, what would be the number of such persons in the general population? The problem may be stated a little more clearly: A certain number of persons out of a marriageable population of 40,000,000 carry defective germ cells. If two of them marry, one-quarter of their children will be feeble-minded. If 100,000 such marriages did occur, what is the ratio of these defect carriers to normals in the general population?

Pairing among defect carriers has occurred in the ratio of 1 to 200 marriages; then these individuals must be present in the general population in the ratio of 1 to 14,[a] if no disturbing factors exist.

The thought that one person out of every fourteen carries the basis of serious mental defectiveness in over half of his or her reproductive cells is enough to make the stoutest heart quake. The problem of cutting off defective germ plasm is not the theoretically simple one of preventing the multiplication of the afflicted; it is the almost hopeless task of reducing the birth rate among the personally unaffected transmitters where there is

---

[a] $\sqrt{\frac{1}{200}}$ = approximately $\frac{13}{14}$. The probability of normal mating normal = $\left(\frac{13}{14}\right)^2 = \frac{169}{196}$, the probability of normal mating carrier is $2\left(\frac{13}{14} \times \frac{1}{14}\right) = \frac{26}{196}$, the probability of two carriers mating is $\left(\frac{1}{14}\right)^2 = \frac{1}{196}$.

little prudential restraint and consequently a high reproductive rate.

The problem exists in just the form we have stated it, but perhaps the picture has been overdrawn. Although the ratio is extremely conservative from one point of view because of the low estimate of defectives and the tremendous birth rate used, it may be considerably too high by reason of our inability to allow for a proper selective marriage rate between the carriers. Goddard, who has made a more intensive study of these persons than anyone else, is of the opinion that the heterozygotes are not in the same class with pure normals. They are more or less dull, stupid, lacking in real ability. For this reason they are unquestionably thrown together more than would otherwise be the case. They tend to form a class of the population which weds within itself. This is not a wholesome thing, but it is much better than having it corrupt the good germ plasm of the country. Although one can make no true estimate, such a selective marriage rate may be high enough to revise our ratio two or three hundred per cent. Instead of 1 out of 14 being of this type, it may be only 1 out of 28 or 42. At best it is food for thought.

Enough has been said about the effect of inbreeding in man to show why the numerous statistical investigations on marriages of near kin have reached no concordant results. Hundreds of such investigations have been made. The earlier ones were compiled by Huth,[97] who came very near the truth considering the state of knowledge of his time. The later data have been brought together by Westermarck,[214] but Westermarck was so imbued with preconceived ideas of what ought to be true that he made matters more chaotic than ever. The impossibility of a correct statistical answer to the problem is clear if one works back from the answer given by the research on heredity: "Inbreeding is not in itself harmful; whatever effect it may have is due wholly to the inheritance received." It is not to be wondered, therefore, that examination of the pedigree record of one family led to one conclusion, and of another family to exactly the opposite.

Nothing has been mentioned about the effect of crossing in the human race, whether such crosses be narrow or wide. Such a discussion belongs more properly in the concluding chapter as it concerns the race more than the individual. Physically outcrossing may often be of value to the individual, but there are reasons to be discussed later for not generalizing hastily in the matter. What we wish to say in conclusion here refers still to inbreeding. It is this:

Owing to the existence of serious recessive traits there is objection to indiscriminate, irrational, intensive inbreeding in man; yet inbreeding is the surest means of establishing families which as a whole are of high value to the community. On the other hand, owing to the complex nature of the mental traits of the highest type, the brightest examples of inherent ability have come and will come from chance mating in the general population, the common people so-called, because of the variability there existent. There can be no permanent aristocracy of brains, because families, no matter how inbred, will remain variable while in existence and will persist but a comparatively short time as close-bred strains. But he is a trifler with little thought of his duty to the state or to himself, who, having ability as a personal endowment, does not scan with care the genealogical record of the family into which he enters.

CHAPTER XIII

THE INTERMINGLING OF RACES AND NATIONAL STAMINA

A PROPOSAL to discuss racial mixtures as a final topic in such a condensed treatment of inbreeding and outbreeding as is presented here, may be deemed somewhat presumptuous, both because of the intricacy and difficulty of the subject itself and because of the immense amount of partially codified knowledge relating to such matters which has been gathered by anthropologists. On the contrary, these very facts are the logical reasons for venturing to indicate how and where the conclusions of experimental genetics can be applied to the problems known collectively as race problems.

The data of anthropology are largely those of the historical type in which the control of variables is always uncertain and often impossible. It is obvious that the nature of the material to a certain extent limits direct investigation in this field to the historical and the statistical methods of research. Generally speaking, one cannot use man as the subject of quantitative laboratory experiments. Yet the difficulties involved demand varied methods of attack; and since genetics has furnished a satisfactory interpretation of heredity by dealing with the lower organisms and has proved that the same mode of inheritance prevails in man, it is inexcusable if the broad ethnological application of the results is neglected. Of course, one must not expect the impossible. Problems so complex can have no genetical solutions permitting predictions in individual cases, but the principles do give information based upon the law of averages which has some importance.

Man is a single species if one may judge from the interfertility and the blood chemistry of existing peoples, but mankind are not all brothers in spite of this oft-repeated laconicism of idealists and radicals; through some 300,000 years of evolution the relationship between the extremes is rather vague. During this period the black race, the yellow race, the white race—three well-marked varieties of the species—have come into existence; and the total number of heritable variations differentiating sub-races and individuals is almost incalculable. Naturally, the selective agents concerned in this process of segregation were numerous; but isolation in a broad sense, necessitating as it did a variety of group struggles for existence amid different environments, probably may be regarded as the factor chiefly responsible. In the immediate past, however, a short period as evolutionary time is marked, there has been an increasingly swift reversal of this process of racial separation. History, in fact, has been hardly more than a record of successive race migrations with the inevitable mingling of the conqueror with the conquered.

With the twentieth century the world enters a new phase of development. Within a single generation man has reached out and grasped the mastery of his environment. Space has been annihilated with the telegraph and telephone, the railway, the steamboat, the submarine, the aëroplane. As a result of this freedom of communication there will be even more colonization until the limit of the food supply is reached; and then, a stationary population, through an increased death rate or a decreased birth rate.

It is unfortunate, in view of the facts in the case, that many should still scoff at the conclusions of Malthus on the subject of population, reached a century ago. The impossibility of the food supply keeping pace with an unchecked natural increase of population is a truism which cannot be glossed over by pointing to the ingenuity of man in applying mechanics to agriculture. The truth is that the world is approaching a population limit faster even than Malthus supposed, and the result of applying new methods to field culture is merely to exploit the natural fertility of the soil at a higher rate. The supposed increase in the amount of food is illusory. In the United States, naturally the richest country on the globe, *the per capita production* of all the important meat animals and some of the great agricultural crops is *decreasing*.

At present the situation is this: China, having reached the limit of her food supply, and having little or no foreign trade, has become stationary in population. Large portions of Europe and the country of Japan have reached the limit of sustenance within themselves, but are increasing at a rate of from ten to fifteen per thousand annually because their commerce is such as to permit importation to supply the deficit. Australia and New Zealand and other parts of Asia and Europe are increasing at a rate which neither their agriculture nor their commerce can long sustain. The Americas and Africa are left as the great centres of colonization. Each will support a large additional number of people; but when they have reached their limit, and that limit will come within a very few centuries—three at most—each country, or at least each continent, must support its own population.

As an outcome of these conditions, the world faces increasing amounts of race amalgamation, and there is naturally an acute interest in race problems. The greater part of this interest is due to prejudice arising from racial and national arrogance. Normally each sub-race believes implicitly in its own superiority and hopes for continued increase and ultimate survival. Perhaps such prejudice prevents any wholly objective discussion of the matter. But apart from desires and hopes concerning racial domination, it ought to be possible to set forth the facts as they are and to determine roughly what ought to occur under given conditions.

In order that there shall be no misunderstanding in regard to the premises taken, let us first consider the classification of man from the anthropological and from the genetic viewpoints.

Anthropologists have been confronted with the very difficult task of classifying existent peoples both with the view of furnishing a useful nomenclature and for the purpose of solving problems of descent. They have recognized the insubstantial character of a language or a nationality basis and have founded their systems on physical traits. Even these, head form, hair shape, skin color, stature, and so on, have been freely acknowledged to be less satisfactory than might be desired. Nevertheless the systems in vogue have been serviceable in many ways, and it is only when they are used as quantitative measures of ancestry that the geneticist is inclined to raise certain objections.

The difficulties of the anthropologist are relatively much greater than those of other systematists. Intergroup sterility is a great aid to botanists and zoölogists.

In general *their* taxonomists have had only to differentiate strains which do not interbreed. The mission of the ethnologist may be compared rather to that of the agriculturist who is called upon to produce a usable classification of the numerous strains of a variable domesticated species, such as cattle and swine, or even wheat and maize. He must for the sake of convenience make a morphological grouping that is non-existent in physical fact. He does this by taking advantage of isolation; without isolation it is impossible.

An appreciation of Mendelian inheritance shows the fallacy involved in making such a system a basis for tracing ethnic relationships. An immense number of hereditary variations have occurred in man. Some can be described by the main structural changes they effect, some cannot. Some distinct changes, such as eye color, have a very simple method of inheritance. They are the mark of single-factor differences in the germ plasm. Other changes, such as those expressed in stature and skull form, appear to be controlled by numerous factors. There are even numerous factor changes which seem to produce no visible effect on the individual and whose existence can be shown only by crossing. For example, it may be assumed with considerable confidence that individuals can have the same cephalic index and yet differ by several hereditary factors whose chief functions are the control of this character. At least such cases have been found in other species and there is no reason for supposing they do not occur in man.

Now these various physical, physiological and psychic characters are controlled by factors transmitted alternatively. They may be linked in various manners, it is true, but they are presumably Mendelian. Consequently one must be very cautious about drawing genetic conclusions in the human race based upon the possession of particular traits, in the absence of proof of a long-continued isolation. Long isolation, it must be assumed, aided in segregating some well-marked human subspecies. It may serve a purpose to continue to accept certain of these types as implied in the terms, white, yellow and black races. Yet one must not forget that real isolation belongs to past epochs. There has been no small amount of interbreeding between even these main types, and the magnitude of the interbreeding between sub-races is largely a matter of historical record. Traits originally characteristic of certain peoples because of isolation and the consequent inbreeding have been shifted back and forth, combined and recombined. It is positively misleading, therefore, to classify Englishmen as resembling Danish, Norman, Pictic, Celtic or Bronze Age types, as is done in more than one work of authority. Even if it were known what the average values of the various characters of these early strains were, there is little reason for believing that a present-day individual bearing one or two particularly striking traits should be felt to hold any closer relationship to the strain in which these traits are supposed to have arisen than his neighbors who are without them. He may have outstanding characters which were once peculiar to a comparatively pure race; but he probably carries these characters as a mere matter of Mendelian recombination. It is wholly possible, for example, that a tall, blue-eyed, dolichocephalic Frenchman really possesses less of the so-called Nordic factors than a short, dark-eyed round-head.

One other matter to be kept in mind in this connection is the impossibility of knowing what factors have survived and what have perished. The great differences between individuals in inherent traits both physical and mental, make it probable that even within a race the average capacity of some strains is greater than others. This seems a fair deduction after making all due allowances for changes in the spirit of the times which accelerate or retard the development of natural ability. Now we do not know and cannot know how the hereditary factors existent to-day compare with those existent 2000 years ago. Selection within the population is an invariable concomitant of human existence. There is a selective death rate, selective mating, selective fertility, each influenced by many conditions. These selective agencies do not remain the same, nor does the material upon which they work. We do not know, for example, whether the most desirable germ plasm of Greece, of Rome, of Mediæval Europe, has been passed on or has ceased to exist. The world has received a great legacy in the creative production of the master men of the past; that it is their heir in physical fact is not so certain. In the strictly biological sense, *i.e.*, in the material basis of heredity, the world may be better or it may be worse than it was in the time of Pericles. This phase of the subject is mentioned because one often hears comments on the degeneracy of certain nations who have had periods of enviable greatness. Social conditions may be the cause, but it should not be forgotten that they are not the sole possible cause. Essential hereditary factors may have been cut off, may have been wholly eliminated.

These genetic ideas of race heredity among mankind

are, we believe, fundamental. They give a clue as to what has happened in the racial mixtures of the past, and enable one to visualize more clearly the probable result of the intense race amalgamation to be expected in the twentieth century.

The world faces two types of racial combination: one in which the races are so far apart as to make hybridization a real breaking-down of the inherent characteristics of each; the other, where fewer differences present only the possibility of a somewhat greater variability as a desirable basis for selection. Roughly, the former is the color-line problem; the latter is that of the White Melting Pot, faced particularly by Europe, North America and Australia.

The genetics of these two kinds of racial intermixture is as follows: Consider first a cross between two extremes, typical members of the white and of the black race. In the first generation the individuals show a notable amount of heterosis, indicating differences in a large number of hereditary factors. They are intermediate in hair form, skin color, head shape, and various other physical attributes, in mental capacity, and in psychical characters in general; although they show extraordinary physical vigor. In later generations segregation and recombination in many of these characters can be traced with little difficulty; but if one describes the descendants of the cross as a population, or even the total characteristics of a single individual, fluctuation around the average of the two original races is still the rule. There may be an approach to the head form of one race combined with the skin color of the other, an approximation of the hair of the one coupled with the other's stature; nevertheless, there is little likelihood of an individual return to the pure type of either race. The difficulties involved are those described in Chapter VII. The races differ by so many transmissible factors, factors which are probably linked in varied ways, that there is, practically speaking, no reasonable chance of such breaks in linkage occurring as would bring together only the most desirable features, even supposing conscious selection could be made. And selection is not conscious. Breeding for the most part is at random. The real result of such a wide racial cross, therefore, is to break apart those compatible physical and mental qualities which have established a smoothly operating whole in each race by hundreds of generations of natural selection.

If the two races possessed equivalent physical characteristics and mental capacities, there would still be this valid genetical objection to crossing, as one may readily see. But in reality the negro is inferior to the white.[181] This is not hypothesis or supposition; it is a crude statement of actual fact. The negro has given the world no original contribution of high merit. By his own initiative in his original habitat, he has never risen. Transplanted to a new environment, as in the case of Haiti, he has done no better. In competition with the white race, he has failed to approach its standard. But because he has failed to equal the white man's ability, his natural increase is low in comparison. The native population of Africa is increasing very slowly, if at all. In the best environment to which he has been subjected, the United States, his ratio in the general population is decreasing. His only chance for an extended survival is amalgamation.

The United States has been confronted by this grave question for some time. In Africa it has hardly yet come to the fore, but within three generations it will be recognized as *the* political and economic problem. What the solution will be, no one knows. It seems an unnecessary accompaniment to humane treatment, an illogical extension of altruism, however, to seek to elevate the black race at the cost of lowering the white. And the statement is made with all due regard to the fact that there are certain desirable characteristics existent in the black race, and that unquestionably the two races overlap in general inherent capacity. The white race as a whole is not equal to the black race in resistance to several serious diseases, as the medical records of the United States army show. The two strains have built up disease resistance along different lines, and the addition of both sets of immunity factors would be desirable. But the practical attainment of such a benefit, given the genetic premises, is so improbable as to be negligible, apart from other considerations.

What would be the result of racial intermixture between the yellow and the white is not so certain. Both races have produced high types. Can one say that either is on the whole the better? The Chinese, the development of very early tribal mixtures, have had a great productive period. In a sense their productivity has decreased, yet their germ plasm is unquestionably good. The Japanese, the result of a much later racial amalgamation, have developed into a wonderful people. Whether it is fair to say the white race is the greater because in the past two centuries they have made such wonderful contributions to civilization is a question. The contributions of the yellow race 4000 years ago were as marvellous in their time. Yellow-white amalgamation may not be fraught with the evil consequences in the wake of the yellow-black and the white-black crosses. At the same time it should be pointed out that the Caucasian and the Mongolian are far apart in descent, and the advantages to be gained by either in thus breaking up superior hereditary complexes developed during an extended past are not clear. At any rate, there seems to be no prospective benefit to the superior yellow peoples in mating with some of the inferior existent whites, and no presumable good to the superior white in intermingling with the poorer yellow offshoots, as has been done to the south of the United States.

Our first conclusion may be said to be a decision against the union of races having markedly different characteristics—particularly when one is decidedly the inferior. Through the operation of the laws of heredity such unions tend to break apart series of character complexes which through years of selection have proved to be compatible with each other and with the persistence of the race under the environment to which it has been subjected. Because of the transmission of factors in linked groups, the low probability of obtaining a single recombination equal or superior to the average of the better race does not warrant the production of multitudes of racial mediocrities which such a mixture entails.

Our second thesis is seemingly paradoxical. It asserts that the foundation stocks of races which have impressed civilization most deeply have been produced by intermingling peoples who through one cause or other became genetically *somewhat* unlike. Theoretically, this theorem is not difficult to develop. Whatever the causes of racial separation under the isolation characteristic of former times, peoples did come to have a rather narrow variability. They were, one might say, homozygous for certain traits. These traits naturally differed in their value. There were great peoples, mediocre peoples, and wretched peoples. But each was more or less standardized. When there came occasion for these standardized peoples differing in their transmissible characters to intermingle, great variability was produced; and if the differences were not too great, the chances were high that valuable character combinations would come to light. Later through the close breeding due to the marriage selection which always develops within human society, the tendency was again to produce purer strains characterized differently; but without the chance of repeated Mendelian recombinations the probability of establishing superior strains was small.

This hypothesis, developed wholly from a consideration of the genetic facts, is not refuted by ethnological data. Thus, if one considers the peoples of Europe, he finds high civilizations, invariably following the migration of that ancient race or mixture of races termed Aryan, a people of whom there is now only circumstantial evidence. It was manifestly not mere hybridization which brought results of outstanding value, however, but hybridization of good strains not too widely differentiated, followed by periods of more or less intensive inbreeding. This is a reasonable deduction from the rapidity with which European and particularly North European culture has outstripped that of Central and Southern Asia.

INTERMINGLING OF RACES

The difficulty with using these data as actual support of the hypothesis under consideration comes from the fact that the amount of hybridization appears to be about the same in various peoples who differ greatly in their contributions to civilization. This may mean that the strains supposedly lower in ability have potentialities not yet realized, or it may mean inherent differences in the original constituent parts. This much seems to be true, however. The great individuals of Europe, the leaders in thought, have come in greater numbers from peoples having very large amounts of ethnic mixture. Even the Scandinavians, a relatively pure strain of the stock to which much of the greatness of Germany, England and even France is supposed to be due, have been somewhat behind these peoples in the production of constructive leaders. Is it not a fair assumption that the backwardness of Spain and Ireland is due to their relative isolation? Is it not because the waves of migration were nearly spent before they reached these lands' ends?

Contrast the people in the United Kingdom, more particularly the natives of the south and west of Ireland, with those of Scotland and England. In proportion to their numbers no modern people has approached the English and Scotch in number of illustrious men or in height of creative ability except the French; the true Irish have hardly a single individual meriting a rank among the great names of history, or a contribution to literature, art, or science of first magnitude.

The Irish are supposed to have arisen somewhat as follows (Ripley,[183] MacNamara[133]). In the early quaternary period, western Europe and northern Africa were occupied by an extremely low type of being of Mongolian antecedents, the Iberian race. Until the neolithic period these tribes were the only inhabitants of the British Isles. During the Stone Age, however, there is evidence of the presence of a second Mongoloid race, the Turanian. Just before the Bronze Age an Aryan stock, the Celts, invaded Britain and Ireland. These people came from the south —France or Spain. Probably they were originally close relatives of the Aryans who migrated from Asia to the northwest and by intermingling with the natives and developing as they went, formed the vigorous Teutonic Aryan or Nordic. But the southern migration of Aryans met very different tribes on their journey, producing in the Celts a somewhat inferior stock. However this may be, the original Celtic horde probably did not make a great impression on the racial character of the Irish; something which also may be said of the second Celtic strain, more highly civilized and warlike than the original visitors, which entered Ireland during the Bronze Age. This later stream of invasion continued over a long period for the island was not completely subjugated until well into the fifth century; but the intruders came as conquerors of a higher social order whose social ideal was to keep their stock uncontaminated with the blood of the native race.

The Norsemen, Nordic Aryans, attempted many times to gain possession of Ireland between the ninth and the fourteenth centuries, but were unsuccessful, and as the Romans and the Saxons never attempted to invade Ireland, the land won by the Celtic chiefs remained in the hands of their direct descendants until 1654, when Cromwell confiscated it, and either killed or reduced them to the condition of laborers.

The present inhabitants of Ireland, then, with the exception of the northern counties, where there is a considerable proportion of English and Scotch, are in the main descended from two savage tribes, the Iberian and Turanian, both probably Mongolian admixtures, with the addition of some blood of the conquering and ruling Celtic Aryans, who genetically must have been more or less intercrossed with Iberian and Turanian tribes by the time they reached the island. Comparatively close interbreeding for at least ten centuries has produced the Irish of to-day.

The original population of Britain, as of Ireland, was Iberian overlaid with Turanian in the north and some other Mongoloid tribes in the south. These races and the types they produced by intermarriage formed the bulk of the population even up to the time of Cæsar's invasion, though the ruling classes were probably Celtic Aryans. The Romans, anthropologists tell us, made little change in the racial character of the inhabitants; but this statement must be taken with some reservation. It is hardly likely that the large garrisons kept by the Roman Empire for 500 years left no descendants. As a population, perhaps, the racial characteristics were not changed to a noticeable degree; nevertheless, a comparatively few thousand persons with Roman blood may have had some considerable effect on the nation as individuals, and the probable presence of this germ plasm must not be counted as negligible. However this may be, the matter is perhaps of little importance as far as the Scotch and English of to-day are concerned, for the greater part of these early peoples, as well as the descendants of the Jutes who entered the country in the fifth century, were exterminated by the Nordic Aryans that invaded the country under the various names of Saxons, Angles and Franks between the sixth and tenth centuries. There was no great racial change made, then, when the Danes conquered the country in the early part of the eleventh century, or when William the Conqueror brought over his Normans of the same stock in the latter part.

The main point we wish to bring out is that England and Scotland are to-day inhabited by an extremely variable people, made so by innumerable crosses into which entered the blood of many Nordic Aryans who differed from each other in some degree. It makes no difference whether there is some variance among ethnologists as to the exactitude of the racial history. That is not essential and one need not quibble about it. The fact remains that the English and Scotch have a generally high civic value and are extremely variable. They produce genius and they produce wretchedness as the natural result of the recombination of these variations. Selections made from the best of these segregates have given the United States names of which one may well be proud; selections made from the other extreme have furnished several of the undesirable strains described previously under the pseudonyms Nam, Juke, etc.

The Irish, on the other hand, and the same might be said of some other isolated types, are much purer from the genetic standpoint. Is there not some reason for attributing to this comparative purity, to this lack of flexibility, their present position as a race and as individuals?

A case similar to that of England and Scotland might be made out of France and for Germany, though France has perhaps a greater proportion of the blood of the

Alpine and Mediterranean peoples than even the south of England and Wales. But even so, the racial differences have not been so great but that France has become one people, with all the chances for good held by a comparatively small united nation, when an amount of close breeding has taken place sufficient to bring out the inherent possibilities.

Another people, great in their influence on the civilization of western Europe, are the Jews. They should not be overlooked in this connection, because of the mistaken idea that they form a pure race of narrow variability characterized by fixed traits. Nothing is further from the truth. If it were the truth it might be questioned whether the Jew would have produced the great number of illustrious men who must in all fairness be credited to them.

The very term race applied to the Jew is a misnomer. There is no more a Jewish race than there is an English race. The fiction has been kept up because of a cult of racial purity in their religion. As a matter of fact, the early Jewish people arose from complex crosses in which at least three different stocks entered: the Arabs, the Assyrioides or Hittites and the Aryan Amorites. More or less inbreeding did follow before their dispersal, but that great racial variability must have remained at the most nationalistic period of their history any student of history knows. After their dispersal there was a period of proselyting which broadened their possibilities. Later, moving into every part of Europe, they mixed with the people with whom they sojourned to a very considerable extent, though keeping up the while the religious ideal of racial purity. In actual fact the Spanish Jew, the German

Jew and the English Jew are to-day scarcely more alike than the Spaniard, the German or the Englishman; but the practical results of their religious beliefs, *since they were attained but partially,* have been good in the genetic sense, for a sufficient amount of inbreeding has prevailed to bring out the possibilities inherent in the combinations made. Civically this conventional isolation has worked to the disadvantage both of the people themselves and of the State in which they held citizenship, and at present it would unquestionably be better from all points of view for them to follow the advice of some of their broader minded leaders in the United States and abandon it, since their own variability has become so great as to make even a theoretically fixed policy of intraracial marriage undesirable. An alliance of a Jew of high capacity and proved worth with a "Nam" or a "Juke" of his own religion is no more to be commended than a similar alliance among those of other faiths.

These three illustrations must suffice as anthropological support of the point we have endeavored to emphasize. In themselves they are not particularly convincing, it must be admitted. Such data can never be used as critical tests of biological theory. At the same time, when considered carefully in the light of the purely genetic facts presented, it seems to us one must assent to the general truth of the theses laid down. Man, like other organic species, has varied markedly in hereditary characters. Races have arisen which are as distinct in mental capacity as in physical traits. These transmissible qualities are governed by germinal factors and these factors are passed on to succeeding generations by the same precise laws that have been discussed in the preceeding pages.

This being true, racial crossing may be desirable or undesirable, depending first on whether the stocks concerned possess a preponderance of desirable characteristics, and second, on whether they are extremely differentiated or not. It may be questioned whether all existing peoples do not possess some desirable traits and hence hold out the possibility of the production of some superior individuals when crossed with presumably superior stock. Nevertheless, even as in breeding for quality in domestic animals, the frequency with which the superior individual is obtained by such a procedure is so low that economically radical experiments are unwise. Given some presumption of equally desirable contribution in the union, the wisdom of a particular racial cross is governed by the number of hereditary differences brought together. The hybridization of extremes is undesirable because of the improbability of regaining the merits of the originals, yet hybridization of somewhat nearly related races is almost a prerequisite to rapid progress, for from such hybridization comes that moderate amount of variability which presents the possibility of the super-individual, the genius.

To produce greatness a nation must have some wretchedness, for such is the law of Mendelian recombination; but the nation that produces wretchedness is not necessarily in the way of producing greatness. There must be racial mixture to induce variability, but these racial crosses must not be too wide else the chances are too few and the time required is too great for the proper recombinations making for inherent capacity to occur. Further, there must be periods of more or less inbreeding following racial mixtures, if there is to be any high probability of isolating desirable extremes. A third essential in the production of racial stamina is that the ingredients in the Melting Pot be sound at the beginning, for one does not improve the amalgam by putting in dross.

May we consider, in conclusion, the bearing of these facts upon the problem of this particular country, the United States of America? The United States at one time was the Mecca of the politically oppressed. Freedom-loving people of good lineage and worthy attainments came to its shores. Now, except for temporary abatement of immigration due to the world war, the stream, though swelling in volume, has changed both its source and the impelling cause of its flow. The early settlers came from stock which had made notable contributions to civilization. They were drawn by a desire from within to carve out great names and fortunes. And they have not disgraced their kin across the seas.

This tide has ebbed, and has been succeeded by a greater. Fifteen million foreign-born now live within the boundaries of the nation, though nearly half have never sought its citizenship. They come in increasing numbers from peoples who have impressed modern civilization but lightly. They come, not so much from inborn ambition of their own, but because attracted by the inducements of those who would exploit them for their own convenience. Whether any considerable part of these people are genetically undesirable, whether real capacity will be discovered under the new environment, no one can say. Time alone will tell. But there is a thought in this connection that cannot be emphasized too strongly or too often. To make this a united nation, there must be an enormous amount of open racial intermixture. The publicist and sociologist should realize that if they do not give their children in marriage with the immigrant, they must with the immigrant's children. Invidious comparisons are, therefore, unnecessary; questions of what this or that race has done or may do need not be settled. It is quite within the province, it is indeed the duty of the native citizen, to require a pause in this mad rush for mere population, until there is a diffusion of education and a healthy growth of a nationalistic spirit. By the time this has been accomplished, the result of the previous policy of the Open Door can be estimated more justly, and any necessary adjustments made with better regard for the good of all the people.

LITERATURE[a]

[1] ALLEN, C. E.: A Chromosome Difference Correlated with Sex Differences. *Science*, N. S., 1917, xlvi, 466, 467.

[2] ARNER, G. B. L.: Consanguineous Marriages in the American Population. *Studies in Hist., Econ. and Pub. Law*, 1909, xxxi, No. 3.

[3] BEAL, W. J.: Reports, Michigan Board of Agriculture, 1876, 1877, 1881 and 1882.

[4] BELL, A. G.: Memoir upon the Formation of a Deaf Variety of the Human Race. *Mem. Nat. Acad. Sci.*, 1884, pp. 86.

[5] BEMISS, S. M.: Report on Influence of Marriages of Consanguinity upon Offspring. *Trans. Amer. Med. Assn.*, 1858, xi, 321–425.

[6] BERTHOLLET, S.: Phénomènes de l'acte mystérieux de la fécondation. *Mém. Soc. Linnéenne de Paris*, 1827, i, 81–83.

[7] BLYTH, E.: On the Physiological Distinctions between Man and all other Animals. *Mag. Nat. His.*, N. S., 1837, i, 1–9, 77–85, 131–141.

[8] BONHOTE, J. L.: Vigour and Heredity. London, 1915, pp. 263.

[9] BOUDIN, M.: Dangers des unions consanguines et nécessité des croisements dans l'espèce humaine et parmi les animaux. *Ann. d'Hygiène pub. et de Méd. légale*, 1862, xviii, 5–82.

[10] BRIDGES, C. B.: Non-disjunction as a Proof of the Chromosome Theory of Heredity. *Genetics*, 1916, i, 1–51, 107–163.

[11] BRIDGES, C. B.: Deficiency. *Genetics*, 1917, ii, 445–465.

[12] BRITTON, E. G.: A Hybrid Moss. *Plant World*, 1898, i, 138.

[13] BRUCE, A. B.: A Mendelian Theory of Heredity and the Augmentation of Vigor. *Science*, N. S., 1910, xxxii, 627, 628.

[14] BRUCE, A. B.: Inbreeding. *Jour. Gen.*, 1917, vi, 195–200.

[15] BURGEOIS, A.: Quelle est l'influence des mariages consanguines sur les générations? *Thèses L'École de Méd.*, 1859, ii, No. 91.

[16] CARRIER, L.: The Immediate Effect of Crossing Strains of Corn. Virginia Agr. Exp. Sta. Bull. 202, 1911, pp. 11.

---

[a] This list of literature makes no pretension of citing other than a few of the most important books and papers on inbreeding published in pre-Mendelian days. Those interested in the subject from the standpoint of marriages of near kin can obtain access to the literature by following up the citations of Huth and of Westermarck. The real development of the subject has come from the investigations on heredity completed since the year 1900. Since it is impracticable and unnecessary to cite all the genetic work of this period, only those titles which are in some way directly connected with the subjects discussed, have been given.

17 CASTLE, W. E.: The Early Embryology of *Ciona intestinalis* Flemming (L.). Mus. Com. Zoöl. Bull. 27, 1896, 201–280.
18 CASTLE, W. E.: Genetics and Eugenics. Cambridge, 1916, pp. 353.
19 CASTLE, W. E., and LITTLE, C. C.: On a Modified Mendelian Ratio Among Yellow Mice. *Science*, N. S., 1910, xxxii, 868–870.
20 CASTLE, W. E., and WRIGHT, S.: Studies of Inheritance in Guinea-pigs and Rabbits. Carnegie Inst. Pub. 241, Washington, 1916, pp. 192.
21 CASTLE, W. E., CARPENTER, F. W. *et al.*: The Effects of Inbreeding, Cross-breeding, and Selection upon the Fertility and Variability of Drosophila. *Proc. Amer. Acad. Arts and Sci.*, 1906, xli, 731–786.
22 CATTELL, J. M.: A Statistical Study of American Men of Science. *Science*, N. S., 1906, xxiv, 658–665, 699–707, 732–742.
23 CAULLERY, M.: Les problèmes de la sexualité. Paris, 1917, pp. 332.
24 CHAMBERLAIN, H. S.: The Foundations of the Nineteenth Century. Trans. J. Lees. 2 vol. New York, 1910, pp. 578–580.
25 CHAPEAUROUGE, A. de: Einiges über Inzucht und ihre Leistung auf verschiedenen Zuchtgebieten. Hamburg, 1909.
26 COLLINS, G. N.: Increased Yields of Corn from Hybrid Seed. Yearbook U. S. Dept. Agr., 1910, 319–328.
27 COLLINS, G. N.: The Value of First Generation Hybrids in Corn. Bull. 191, Bur. Plant Ind., U. S. Dept. Agr., 1910, pp. 45.
28 COLLINS, G. N., and KEMPTON, J. H.: Effects of Cross-pollination on the Size of Seed in Maize. Cir. 124 U. S. Dept. Agr. 1913, 9–15.
29 COLLINS, G. N.: A More Accurate Method of Comparing First Generation Hybrids with Their Parents. *Jour. Agr. Res.*, 1914, iii, 85–91.
30 COLLINS, G. N.: Maize: Its Origin and Relationships. Notes of the 123d Regular Meeting Bot. Soc. Wash., *Jour. Wash. Acad. Sci.*, 1918, viii, 42, 43.
31 COOK, O. F.: The Superiority of Line Breeding Over Narrow Breeding. Bull. 146, U. S. Dept. Agr., Bur. Plant Ind., 1909, pp. 45.
32 COULTER, J. M.: The Evolution of Sex in Plants. Chicago, 1914, pp. 140.
33 CRAMER, P. J. S.: Kritische Uebersicht der bekannten Fälle von Knospenvariation. Haarlem, 1907, pp. 474.
34 CRAMPE, H.: Zuchtversuche mit zahmen Wanderratten. *Landw. Jahrb.*, 1883, xii, 389–458.
35 CULL, S. W.: Rejuvenescence as the Result of Conjugation. *Jour. Exp. Zoöl.*, 1907, iv, 85–89.
36 DAFFNER, F.: Das Wachstum des Menschen. Leipzig, 1902, pp. 475.

[37] DANIELSON, F. H., and DAVENPORT, C. B.: The Hill Folk. Mem. 1, Eugenics Record Office, Cold Spring Harbor, 1912, pp. 56.
[38] DARWIN, C.: The Variation of Animals and Plants Under Domestication. 2nd Ed., London, 1875, 2 vols., pp. 461–478.
[39] DARWIN, C.: The Effects of Cross- and Self-Fertilization in the Vegetable Kingdom. London, 1876, pp. 482.
[40] DAVENPORT, C. B.: Heredity in Relation to Eugenics. New York, 1911, pp. 298.
[41] DAVENPORT, C. B.: Inheritance of Stature. *Genetics*, 1917, ii, 313–389.
[42] DAVENPORT, C. B., and DAVENPORT, G.: Heredity of Eye-color in Man. *Science*, N. S., 1907, xxvi, 589–592.
[43] DAVENPORT, C. B., and DAVENPORT, G.: Heredity of Hair-form in Man. *Amer. Nat.*, 1908, xlii, 341–349.
[44] DAVENPORT, C. B., and DAVENPORT, G.: Heredity of Hair-color in Man. *Amer. Nat.*, 1909, xliii, 193–211.
[45] DAVENPORT, C. B., and DAVENPORT, G.: Heredity of Skin-pigmentation in Man. *Amer. Nat.*, 1910, xliv, 641–672; 705–731.
[46] DENIKER, J.: The Races of Man. New York, 1906, pp. 611.
[47] DETLEFSEN, J. A.: Genetic Studies on a Cavy Species Cross. Carnegie Pub. 205, Washington, 1914, pp. 134.
[48] DUGDALE, R. L.: The Jukes. New York, 1877, 4th Ed., 1910, pp. 121.
[49] DÜSING, K.: Die Factoren welche die Sexualität entscheiden. Jena, 1883. Inaug. Dissertation.
[50] EAST, E. M.: Inbreeding in Corn. Connecticut Agr. Exp. Sta. Rpt. for 1907, 1908, 419–428.
[51] EAST, E. M.: A Study of the Factors Influencing the Improvement of the Potato. Illinois Agr. Exp. Sta. Bull. 127, 1908, 375–456.
[52] EAST, E. M.: The Distinction Between Development and Heredity in Inbreeding. *Amer. Nat.*, 1909, xliii, 173–181.
[53] EAST, E. M.: An Interpretation of Sterility in Certain Plants. *Proc. Amer. Phil. Soc.*, 1915, liv, 70–72.
[54] EAST, E. M.: Studies on Size Inheritance in Nicotiana. *Genetics*, 1916, i, 164–176.
[55] EAST, E. M.: The Bearing of Some General Biological Facts on Bud-variation. *Amer. Nat.*, 1917, li, 129–143.
[56] EAST, E. M.: Hidden Feeble-mindedness. *Jour. Her.*, 1917, viii, 215–217.
[57] EAST, E. M.: The Rôle of Reproduction in Evolution. *Amer. Nat.*, 1918, lii, 273–289.
[58] EAST, E. M., and HAYES, H. K.: Inheritance in Maize. Conn. Agr. Exp. Sta. Bull. 167, 1911, pp. 141.

59 EAST, E. M., and HAYES, H. K.: Heterozygosis in Evolution and in Plant Breeding. Bull. 243, U. S. Dept. Agr., Bur. Plant Ind., 1912, pp. 58.

60 ELLIS, H.: A Study of British Genius. London, 1904, pp. 300.

61 EMERSON, R. A.: Inheritance of Sizes and Shapes in Plants. *Amer. Nat.*, 1910, xliv, 739–746.

62 EMERSON, R. A.: The Inheritance of Certain Abnormalities in Maize. Amer. Breed. Assn. Rpt. 1912, viii, 385–399.

63 EMERSON, R. A., and EAST, E. M.: The Inheritance of Quantitative Characters in Maize. Nebraska Agr. Exp. Sta., Research Bull. 2, 1913, pp. 120.

64 ENRIQUES, P.: La conjugazione e il differenziamento negli Infusoria. *Arch. f. Protistenkünde*, 1907, ix, 195–296.

65 ESTABROOK, A. H., and DAVENPORT, C. B.: The Nam Family. Mem. 2, Eugenics Record Office. Cold Spring Harbor, 1912, pp. 85.

66 FABRE-DOMENGUE, P.: Unions consanguines chez les colombins. *L'Intermédiare des Biol.*, 1898, i, pp. 203.

67 FAY, E. A.: Marriages of the Deaf in America. Washington, 1898, pp. 527.

68 FISCHER, E.: Die Rehobother Bastards und das Bastardierungsproblem beim Menschen. Jena. Review. *Jour. Her.*, 1914, v, 465–468.

69 FISH, H. D.: On the Progressive Increase of Homozygosis in Brother-Sister Matings. *Amer. Nat.*, 1914, xlviii, 759–761.

70 FOCKE, W. O.: Die Pflanzen-Mischlinge. Berlin, 1881, pp. 569.

71 FRAZER, J. S.: Totemism and Exogamy. 4 vol., London, 1910.

72 FREUD, S.: Totem and Taboo. Trans. A. A. Brill. New York, 1918, pp. 265.

73 GALTON, F.: Hereditary Genius. 2nd Ed., London, 1892, pp. 379.

74 GÄRTNER, C. F.: Versuche und Beobachtungen über die Bastardererzeugung im Pflanzenreich. Stuttgart, 1849, pp. 790.

75 GAY, C. W.: The Breeds of Livestock. New York, 1916, pp. 483.

76 GENTRY, N. W.: Inbreeding Berkshires. *Amer. Breed. Assn. Ann. Rpt.*, 1905, i, 168–171.

77 GERNERT, W. B.: Aphis Immunity of Teosinte-Corn Hybrids. *Science*, N. S., 1917, xlvi, 390–392.

78 GERSCHLER, M. W.: Ueber alternative Vererbung bei Kreuzung von Cyprinodontiden-Gattungen. *Zeitschr. f. ind. Abst. u. Vererb.*, 1914, xii, 73–96.

79 GOBINEAU, Le Compte de: Essai sur l'inégalité des races humaines. 2 vol. Paris, 1884, pp. 561–566.

80 GODDARD, H. H.: The Kallikak Family. New York, 1913, pp. 121.

[81] GODDARD, H. H.: Feeble-mindedness: Its Causes and Consequences. New York, 1914, pp. 599.
[82] GOLDSCHMIDT, R.: Zuchtversuche mit Enten, I. *Ztschr. f. ind. Abst. u. Vererb.*, 1913, ix, 161–191.
[83] GOURDON, J.: Consanguinité chez les animaux domestiques. *Ann. d'Hygiène pub. et de Méd. légale*, 1862, xviii, 463, 464.
[84] GRANT, M.: The Passing of the Great Race. 2nd Ed. New York, 1918, pp. 295.
[85] GRAVATT, F.: A Radish-cabbage Hybrid. *Jour. Her.*, 1914, v, 269–272.
[86] GUAITA, G. von: Versuche mit Kreuzungen von verschiedenen Rassen der Hausmaus. *Ber. d. Naturforsch. Gesell. zu Freiburg*, 1898, x, 317–332.
[87] GUAITA, G. von: Zweite Mittheilung über Versuche mit Kreuzungen von verschiedenen Hausmausrassen. *Ber. d. Naturforsch. Gesell. zu Freiburg*, 1900, xi, 131–143.
[88] HADDON, A. C.: Races of Man. London, 1909, pp. 126.
[89] HAMMOND, J.: On Some Factors Controlling Fertility in Domestic Animals. *Jour. Agr. Sci.*, 1914, vi, 263–277.
[90] HARDY, G. H.: Mendelian Proportions in a Mixed Population. *Science*, N. S., 1908, xxviii, 49, 50.
[91] HARTLEY, C. P., et al.: Cross-breeding Corn. Bull. 218, U. S. Dept. Agr., Bur. Plant Ind., 1912, pp. 72.
[92] HAYES, H. K.: Corn Improvement in Connecticut. Connecticut Agr. Exp. Sta., Rpt. for 1913, 1914, 353–384.
[93] HAYES, H. K., and EAST, E. M.: Improvement in Corn. Connecticut Agr. Exp. Sta. Bull. 168, 1911, pp. 21.
[94] HAYES, H. K., and EAST, E. M.: Further Experiments on Inheritance in Maize. Connecticut Agr. Exp. Sta. Bull. 188, 1915, pp. 31.
[95] HAYES, H. K., and JONES, D. F.: The Effects of Cross- and Self-fertilization on Tomatoes. Connecticut Agr. Exp. Sta. Rpt. for 1916, 1917, 305–318.
[96] HERBERT, W.: Amaryllidaceæ. London, 1837, pp. 428.
[97] HUTH, A. H.: The Marriage of Near Kin. London, 1875, pp. 359.
[98] HYDE, R. H.: Fertility and Sterility in *Drosophila ampelophila*. *Jour. Exp. Zoöl.*, 1914, xvii, 141-171, 173–212.
[99] JACOBY, P.: Études sur la sélection chez l'homme. 2nd Ed., Paris, 1904.
[100] JANSSENS, F. A.: La théorie de la chiasmatypie. *La Cellule*, 1909, xxv, 389–414.

101 JENNINGS, H. S.: Heredity, Variation and Evolution in Protozoa. II. Heredity and Variation of Size and Form in Paramecium, with Studies of Growth, Environmental Action, and Selection. *Proc. Amer. Phil. Soc.*, 1908, xlvii, 393-546.

102 JENNINGS, H. S.: Production of Pure Homozygotic Organisms from Heterozygotes by Self-Fertilization. *Amer. Nat.*, 1912, xlvi, 487-491.

103 JENNINGS, H. S.: The Effect of Conjugation in Paramecium. *Jour. Exp. Zoöl.*, 1913, xiv, 279-391.

104 JENNINGS, H. S.: Formulæ for the Results of Inbreeding. *Amer. Nat.*, 1914, xlviii, 693-696.

105 JENNINGS, H. S.: The Numerical Results of Diverse Systems of Breeding. *Genetics*, 1916, i, 53-89.

106 JENNINGS, H. S.: The Numerical Results of Diverse Systems of Breeding, with Respect to Two Pairs of Characters, Linked or Independent, with Special Relation to the Effects of Linkage. *Genetics*, 1917, ii, 97-154.

107 JENNINGS, H. S.: Heredity, Variation and the Results of Selection in the Uniparental Reproduction of *Difflugia corona*. *Genetics*, 1916, i, 407-534.

108 JENNINGS, H. S., and LASHLEY, K. S.: Biparental Inheritance and the Question of Sexuality in Paramecium. *Jour. Exp. Zoöl.*, 1913, xiv, 393-466.

109 JOHANNSEN, W.: Ueber Erblichkeit in Populationen und in reinen Linien. Jena, 1903, pp. 68.

110 JOHANNSEN, W.: Elemente der exakten Erblichkeitslehre. Jena, 1909, pp. 515.

111 JONES, D. F.: Dominance of Linked Factors as a Means of Accounting for Heterosis. *Proc. Nat. Acad. Sci.*, 1917, iii, 310-312. Also *Genetics*, 1917, ii, 466-479.

112 JONES, D. F.: Bearing of Heterosis upon Double Fertilization. *Bot. Gaz.*, 1918, lxv, 324-333.

113 JONES, D. F.: The Effects of Inbreeding and Cross-breeding upon Development. Conn. Agr. Exp. Sta. Bull. 207, 1918, pp. 100.

114 JONES, D. F., and HAYES, H. K.: Increasing the Yield of Corn by Crossing. Connecticut Agr. Exp. Sta. Rpt. for 1916, 1917, 323-347.

115 JÖRGER, J.: Die Familie Zero. *Arch. f. Rass. u. Gesellschaftsbiologie*, 1905, ii, 494-559.

116 KEEBLE, F., and PELLEW, C.: The Mode of Inheritance of Stature and of Time of Flowering in Peas (*Pisum sativum*). *Jour. Gen.*, 1910, i, 47-56.

117 KING, H. D.: On the Normal Sex Ratio and the Size of the Litter in the Albino Rat (*Mus norvegicus albinus*). *Anat. Rec.*, 1915, ix, 403–419.

118 KING, H. D.: The Relation of Age to Fertility in the Rat. *Anat. Rec.*, 1916, xi, 269–287.

119 KING, H. D.: Studies on Inbreeding. I. The Effects of Inbreeding on the Growth and Variability in Body Weight of the Albino Rat. *Jour. Exp. Zoöl.*, 1918, xxvi, 1–54.

120 KING, H. D.: Studies on Inbreeding. II. The Effects of Inbreeding on the Fertility and on the Constitutional Vigor of the Albino Rat. *Jour. Exp. Zoöl.*, 1918, xxvi, 55–98.

121 KING, H. D.: Studies on Inbreeding. III. The Effects of Inbreeding with Selection, on the Sex Ratio of the Albino Rat. *Jour. Exp. Zoöl.*, 1918, xxvii, 1–35.

122 KNIGHT, T. A.: An Account of Some Experiments on the Fecundation of Vegetables. *Phil. Trans. Roy. Soc.*, Lon., 1799, lxxxix, 195–204.

123 KNIGHT, T. A.: Physiological and Horticultural Papers. London, 1841, pp. 389.

124 KNUTH, P.: Handbuch der Blütenbiologie. Leipzig, 1898–1905. 3 vol.

125 KÖLREUTER, J. G.: Dritte Fortsetzung der vorläufigen Nachricht von einigen das Geschlecht der Pflanzen betreffenden Versuchen und Beobactungen. Leipzig, 1766, pp. 156. (Reprinted in Ostwald's Klassiker der exakten Wissenschaften, No. 41, Leipzig, 1893.)

126 KRAEMER, H.: Ueber die ungünstigen Wirkungen naher Inzucht. *Mitt. d. deut. landw. Gesell.*, 6 and 13, 1913. Trans. *Jour. Her.*, 1913, v, 226–234.

127 LECOQ, H.: De la fécondation naturelle et artificielle de végétaux et de l'hybridation. Paris, 1845, pp. 287.

128 LINDLEY, J.: The Theory of Horticulture. 2nd Ed., New York, 1852, pp. 364.

129 LOEB, J.: The Organism as a Whole. From a Physico-chemical Viewpoint. New York, 1916, pp. 379.

130 MARCHAL, ÉL., and MARCHAL ÉM.: Aposporie et sexualité chez les Mousses. I, II, III. *Bull. Acad. Roy. Belg.*, Cl. Sci., 1907, 765–789; 1909, 1249–1288; 1911, 750–778.

131 McCLUER, G. W.: Corn Crossing. Illinois Agr. Exp. Sta. Bull. 21, 1892, 82–101.

132 McCULLOCH, O. C.: The Tribe of Ishmael: A Study in Social Degradation. *Proc. 15th Natl. Conf. Char. and Cor.*, 1888.

LITERATURE

133 MacNamara, N. C.: Origin and Character of the British People. London, 1900, pp. 242.

134 Marshall, F. H. A.: The Physiology of Reproduction. London, 1910, pp. 706.

135 Martin, R.: Lehrbuch der Anthropologie in systematischer Darstellung. Jena, 1914, pp. 1181.

136 Maupas, E.: Recherches expérimentales sur la multiplication des infusoires ciliés. *Arch. d. Zoöl. Exp. et Gén.*, II, 1889, vi, 165–277.

137 Maupas, E.: La rajeunissement karyogamique chez les ciliés. *Arch. d. Zoöl. Exp. et Gén.*, II, 1889, vii, 149–517.

138 Mauz, E.: In Correspondenzblatt des Württemburgischen Landw. Ver. 1825.

139 Mendel, G. J.: Versuche über Pflanzen-Hybriden. *Verh. Naturf. Ver. in Brünn*, 1865. Trans. in *Castle's Genetics and Eugenics*, Cambridge, 1916, pp. 281–321.

140 Merz, J. T.: A History of European Thought in the Nineteenth Century. 3rd Ed., 3 vol., London, 1907.

141 Metz, C. W.: The Linkage of Eight Sex-linked Characters in *Drosophila virilis*. *Genetics*, 1918, 107–134.

142 Middleton, A. R.: Heritable Variations and the Results of Selection in the Fission Rate of *Stylonychia pustulata*. *Jour. Exp. Zoöl.*, 1915, xix, 451–503.

143 Mitchell, A.: Blood-relationship in Marriage, Considered in Its Influence upon the Offspring. *Mem. Anthropol. Soc.*, Lon., 1865, ii, 402–456.

144 Moenkhaus, W. J.: The Effects of Inbreeding and Selection on Fertility, Vigor and Sex-ratio of *Drosophila ampelophila*. *Jour. Morph.*, 1911, xxii, 123–154.

145 Montgomery, E. G.: Preliminary Report on Effect of Close and Broad Breeding on Productiveness in Maize. Nebraska Agr. Exp. Sta., 25th Ann. Rpt., 1912, 181–192.

146 Morgan, T. H.: Sex-limited Inheritance in Drosophila. *Science*, N. S., 1910, xxxii, 120–122.

147 Morgan, T. H.: Chromosomes and Associative Inheritance. *Science*, N. S., 1911, xxxiv, 636–638.

148 Morgan, T. H.: An Attempt to Analyze the Constitution of the Chromosomes on the Basis of Sex-limited Inheritance in Drosophila. *Jour. Exp. Zoöl.*, 1911, xi, 365–413.

149 Morgan, T. H.: Heredity and Sex. New York, 1913, pp. 282.

150 Morgan, T. H., and Bridges, C. B.: Sex-linked Inheritance in Drosophila. Carnegie Ins. Pub. 237, Washington, 1916, pp. 87.

151 MORGAN, T. H., STURTEVANT, A. H., MULLER, H. J., and BRIDGES, C. B.: The Mechanism of Mendelian Heredity. New York, 1915, pp. 262.
152 MORROW, G. E., and GARDNER, F. D.: Field Experiments with Corn. Illinois Agr. Exp. Sta. Bull. 25, 1893, 173–203.
153 MORROW, G. E., and GARDNER, F. D.: Experiments with Corn. Illinois Agr. Exp. Sta. Bull. 31, 1894, 359, 360.
154 MULLER, H. J.: The Mechanism of Crossing Over. I, II, III, IV. Amer. Nat., 1916, l, 193–221, 284–305, 350–366, 421–434.
155 MULLER, H. J.: An Œnothera-like Case in Drosophila. Proc. Nat. Acad. Sci., 1917, iii, 619–626.
156 MULLER, H. J.: Genetic Variability, Twin Hybrids and Constant Hybrids, in a Case of Balanced Lethal Factors. Genetics, 1918, iii, 422–499.
157 MÜLLER, H.: Die Befruchtung der Blumen durch Insekten und die gegenseitigen Anpassungen beider. Leipzig, 1873, pp. 478.
158 MUMFORD, F. B.: The Breeding of Animals. New York, 1917, pp. 303.
159 NAUDIN, C.: Nouvelles recherches sur l'hybridité dans les végétaux. Nouv. Arch. du. Mus. d'Hist. Nat. de Paris, 1865, i, 25–174.
160 NEARING, S.: Geographical Distribution of American Genius. Pop. Sci. Mon., 1914, lxxxv, 189–199.
161 NEARING, S.: The Younger Generation of American Genius. Sci. Mon., 1916, ii, 48–61.
162 NĚMEC, B.: Das Problem der Befruchtungsvorgänge. Berlin, 1910, pp. 532.
163 ODIN, A.: Génése des grandes hommes gens de lettres français modernes. 2 vol., Paris, 1895.
164 PARKER, G. H., and BULLARD, C.: On the Size of Litters and the Number of Nipples in Swine. Proc. Amer. Acad. Arts and Sci., 1913, xlix, 397–426.
165 PEARL, R.: The Mode of Inheritance of Fecundity in the Domestic Fowl. Jour. Exp. Zoöl., 1912, xiii, 153–268.
166 PEARL, R.: On the Correlation Between the Number of Mammæ of the Dam and Size of Litter in Mammals. I. Interracial Correlation. Proc. Soc. Exp. Biol. and Med., 1913, xi, 27–30.
167 PEARL, R.: On the Correlation Between the Number of Mammæ of the Dam and Size of Litter in Mammals. II. Intraracial Correlation in Swine. Proc. Soc. Exp. Biol. and Med., 1913, xi, 31, 32.
168 PEARL, R.: A Contribution Towards an Analysis of the Problem of Inbreeding. Amer. Nat., 1913, xlvii, 577–614.

169 PEARL, R., and MINER, J. R.: Tables for Calculating Coefficients of Inbreeding. Maine Agr. Exp. Sta. Rept. for 1913, 191–202.

170 PEARL, R.: On the Results of Inbreeding a Mendelian Population; a Correction and Extension of Previous Conclusions. *Amer. Nat.*, 1914, xlviii, 57–62.

171 PEARL, R.: On a General Formula for the Constitution of the N'th Generation of a Mendelian Population in which all Matings are of Brother $\times$ Sister. *Amer. Nat.*, 1914, xlviii, 491–494.

172 PEARL, R.: Inbreeding and Relationship Coefficients. *Amer. Nat.*, 1914, xlviii, 513–523.

173 PEARL, R.: Modes of Research in Genetics. New York, 1915, pp. 182.

174 PEARL, R.: Some Further Considerations Regarding Cousin and Related Kinds of Mating. *Amer. Nat.*, 1915, xlix, 570–575.

175 PEARL, R.: Some Further Considerations Regarding the Measurement and Numerical Expression of Degrees of Kinship. *Amer. Nat.*, 1917, li, 545–559.

176 PEARL, R.: A Single Numerical Measure of the Total Amount of Inbreeding. *Amer. Nat.*, 1917, li, 636–639.

177 PEARSON, K.: On a Generalized Theory of Alternative Inheritance, with Special References to Mendel's Laws. *Phil. Trans. Roy. Soc.* (A), 1904, cciii, 53–86.

178 PHILLIPS, J. C.: Size Inheritance in Ducks. *Jour. Exp. Zoöl.*, 1912, xii, 369–380.

179 PHILLIPS, J. C.: A Further Study of Size Inheritance in Ducks, with Observations on the Sex Ratio of Hybrid Birds. *Jour. Exp. Zoöl.*, 1914, xvi, 131–148.

180 PLATE, L.: Vererbungslehre. Leipzig, 1913, pp. 519.

181 POPENOE, P., and JOHNSON, R. H.: Applied Eugenics. New York, 1918, pp. 459.

182 PUNNETT, R. C., and BAILEY, P. G.: On Inheritance of Weight in Poultry. *Jour. Gen.*, 1914, iv, 23–39.

183 RIPLEY, W. Z.: The Races of Europe. New York, 1899, pp. 624.

184 RITZEMA-BOS, J.: Untersuchungen über die Folgen der Zucht in engster Blutverwandtschaft. *Biol. Cent.*, 1894, xiv, 75–81.

185 ROBERTS, H. F.: First Generation Hybrids of American and Chinese Corn. Amer. Breed. Assn. Rpt. 1912, viii, 367–384.

186 ROBBINS, R. B.: Some Applications of Mathematics to Breeding Problems. I, II, III. *Genetics*, 1917, ii, 489–504; 1918, iii, 73–92; 1918, iii, 375–389.

187 ROBBINS, R. B.: Random Mating with the Exception of Sister by Brother Mating. *Genetics*, 1918, iii, 390–396.

188 ROMMELL, G. M.: The Fecundity of Poland-China and Duroc-Jersey Sows. Cir. 95, U. S. Dept. Agr., Bur. An. Ind., 1906, pp. 12.

189 ROMMELL, G. M.: The Inheritance of Size of Litter in Poland-China Sows. Amer. Breed. Assn. Rpt., 1907, v, 201–208.

190 ROMMELL, G. M., and PHILLIPS, E. F.: Inheritance in the Female Line of Size of Litter in Poland-China Sows. *Proc. Amer. Phil. Soc.*, 1906, xlv, 245–254.

191 SAGERET, A.: Considérations sur la production des hybrides, des variantes et des variétés en général, et sur celles de la famille de Cucurbitacées en particulier *Ann. des Sci. Nat.*, 1826, viii, 294–314.

192 SHAMEL, A. D.: The Effects of Inbreeding in Plants. Yearbook U. S. Dept. Agr., Washington, 1905, pp. 377–392.

193 SHULL, G. H.: The Composition of a Field of Maize. Amer. Breed. Assn. Rpt., 1908, iv, 296–301.

194 SHULL, G. H.: A Pure Line Method of Corn Breeding. Amer. Breed. Assn. Rpt., 1909, v, 51–59.

195 SHULL, G. H.: Hybridization Methods in Corn Breeding. *Amer. Breed. Mag.*, 1910, i, 98–107.

196 SHULL, G. H.: The Genotypes of Maize. *Amer. Nat.*, 1911, xlv, 234–252.

197 SHULL, G. H.: Duplicate Genes for Capsule Form in *Bursa bursa-pastoris*. *Zeitschr. f. ind. Abst. u. Vererb.*, 1914, xii, 97–149.

198 SHULL, A. F.: The Influence of Inbreeding on Vigor in *Hydatina senta*. *Biol. Bull.*, 1912, xxiv, 1–13.

199 SHULL, A. F.: Studies in the Life Cycle of *Hydatina senta*. III. *Jour. Exp. Zoöl.*, 1912, xii, 283–317.

200 STRASBURGER, E.: Versuche mit diöcischen Pflanzen in Rücksicht auf Geschlechtsverteilung. *Biol. Centralbl.*, 1900, xx, 657.

201 STRASBURGER, E.: Ueber geschlechtbestimmende Ursachen. *Jahrb. Wiss. Bot.*, 1910, xlviii, 427–520.

202 STURTEVANT, A. H.: The Linear Arrangement of Six Sex-linked Factors in Drosophila, as Shown by Their Mode of Association. *Jour. Exp. Zoöl.*, 1913, xiv, 43–59.

203 STURTEVANT, A. H.: The Behavior of the Chromosomes as Studied through Linkage. *Zeitschr. f. ind. Abstam. u. Vererb.*, 1915, xiii, 234–287.

204 SURFACE, F. M.: Fecundity in Swine. *Biometrika*, 1909, vi, 433–436.

205 TOYAMA, K.: Mendel's Law of Heredity as Applied to Silkworm Crosses. *Biol. Chl.*, 1906, xxvi, 321–334.

206 VOISIN, A.: Contribution à l'histoire des mariages entre consanguins. *Compt. rend. Acad. Sci.*, 1865, lxv, 105–108.

207 WARREN, H. C.: Numerical Effects of Natural Selection Acting Upon Mendelian Characters. *Genetics*, 1917, ii, 305–312.
208 WEINSTEIN, A.: Coincidence of Crossing Over in *Drosophila melanogaster* (*ampelophila*). *Genetics*, 1918, iii, 135–159.
209 WEISMANN, A. The Evolution Theory. (Trans. J. A. Thomson and M. R. Thomson.) London, 1904, 2 vol.
210 WELLINGTON, R.: Influence of Crossing in Increasing the Yield of the Tomato. New York Agr. Exp. Sta. Bull. 346, 1912, 57–76.
211 WENTWORTH, E. N.: The Segregation of Fecundity Factors in Drosophila. *Jour. Gen.*, 1913, iii, 113–120.
212 WENTWORTH, E. N., and AUBEL, C. E.: Inheritance of Fertility in Swine. *Jour. Agr. Res.*, 1916, v, 1145–1160.
213 WENTWORTH, E. N., and REMICK, B. L.: Some Breeding Properties of the Generalized Mendelian Population. *Genetics*, 1916, i, 608–616.
214 WESTERMARCK, E.: The History of Human Marriage. 3rd Ed., London, 1903, pp. 644.
215 WHEELER, W. M.: The Ants of the Baltic Amber. *Schrift. Physikökonom. Gesell. Königsberg*, 1914, lv, pp. 142.
216 WHITNEY, D. D.: Reinvigoration Produced by Cross-fertilization in *Hydatina senta*. *Jour. Exp. Zoöl.*, 1912, xii, 337–362.
217 WHITNEY, D. D.: "Strains" in Hydatina. *Biol. Bull.*, 1912, xxii, 205–218.
218 WHITNEY, D. D.: Weak Parthenogenetic Races of *Hydatina senta* Subjected to a Varied Environment. *Biol. Bull.*, 1912, xxiii, 321–330.
219 WIEGMANN, A. F.: Ueber die Bastarderzeugung im Pflanzenreich. Braunschweig, 1828, pp. 40.
220 WITHINGTON, C. F.: Consanguineous Marriages: Their Effect upon Offspring. *Mass. Med. Soc.*, 1885, xiii, 453–484.
221 WOLFE, T. K.: Further Evidence of the Immediate Effect of Crossing Varieties of Corn on the Size of the Seed Produced. *Jour. Amer. Soc. Agr.*, 1915, vii, 265–272.
222 WOODRUFF, L. L.: Two Thousand Generations of Paramecium. *Arch. f. Protistenkunde*, 1911, xxi, 263–266.
223 WOODS, F. A.: Heredity in Royalty. New York, 1906, pp. 312.
224 WOODS, F. A.: The Influence of Monarchs. New York, 1913, pp. 421.
225 WRIGHT, S.: The Effects of Inbreeding on Guinea-pigs. I. Decline in Vigor. II. Differentiation among Inbred Families. III. Crosses between Different Highly Inbred Families. (Doctor Wright kindly permitted the authors to read these valuable unpublished papers in manuscript.)

INDEX

Achondroplasy, 230
Adaptation, for cross-pollination, 34
   for self-pollination, 30
Africa, 247, 254, 257
   native population of, 253
Agassiz, 236
Agriculture, 18
Algæ, 204
Allelomorphs, 55
Allen, 45
Alternation of generations, 29, 46
Althæa, 144
America, 247, 252
Amœba in division, 21
Amorites, 261
Amphimixis, 206
Angles, 260
Annelids, 21
Anthropology, 18, 245
Apomixis, 32
Apple, 210
Arabs, 261
Armadillo, 42
   nine-banded, identical quadruplets in, 44
Arthropoda, 22
Arthropods, 21*ff.*, 207
Aryan, 258
   Celtic, 259
   Nordic, 259*f.*
   races, 256
   Teutonic, 258
Asia, 247, 258
   culture of, 256
Assyrioides, 261
Atavism, 166
Ataxia, Friedrich's, 231
Australia, 247, 252
Autogamy, 30

Barley, 114, 210
Basidiomycetes, 31

Bateson, 165
Beal, 221
Beans, 114
Bees, 158
Berthollet, 141
Bibos, frontalis, 192
   gaurus, 192
   gruniens, 192
Birds, 158
Bison americanus, 192
Blossom's Glorene, 85*ff.*
Bos taurus, 192
Brachydactyly, 230
Brassica oleracea, 192
Breckenridge, 235
Bridges, 184
Britain, 258*f.*
British Isles, 258
Bronze Age, 258
Brünnow, 236
Bryozoans, 23
Budin, 227
Buffalo, 192

Cabbage, 192
Cæsar's invasion of Britain, 259
Calceolaria, 155
Castle, 25, 111, 137, 158, 160, 188, 236
Cat, 211
Catalpa, 150
Cataract, 230
Cattell, 233
Cattle, 210, 213
Caucasian, 255
Caullery, 22
Cavia, 103
   cutleri, 160
   species, hybrids, 192
Celts, 258
Cereals, 211
Cephalic index, 249

Chemotropism, 154
China, 247
Chinese, 254
Chordates, 21
Chromosomes, 36
Ciona intestinalis, 25
Cirripedes, 25, 34
Cleft palate, 230
Coccinea, 144
Coefficient of cross-relationship, 86
   of heredity, 196, 199
   of inbreeding, 81*ff*.
   of relationship, 81, 84*ff*.
Cœlenterates, 21
Collins, 137, 153
Coloration, protective, 146
Color-blindness, 47*ff*.
Complemental males, 25
Compositæ, 32
Conjugatæ, 27
Connate seeds of maize, 133
Consanguinity, 113, 139
Corn, varieties of, 215
Coulter, 26
Cow, 192
Cramer, 198
Crampe, 101*ff*.
Cromwell, 258
Crossing-over, diagram to illustrate, 64
Cucumber, 221
Cucumis, 144
Cymothoidæ, 24
Cytoplasm of egg, 200

Dalton, 235
Danes, 260
Darwin, 13, 25, 32, 34, 101, 114*ff*., 137*ff*., 143, 146*ff*., 154, 164*f*., 186, 235
Datura, 142, 144
Davenport, 230*f*., 233
Delphino, 34
Detlefsen, 103, 192
Diabetes insipidus, 230
Dianthus, 116*f*., 142, 144, 155

Diatomeæ, 27
Dichogamy, 33
Difflugia, 203
   coronata, 78
Digitalis, 144, 155
Diœcism, 22
Disease, susceptibility and resistance to, 134
Dog, 210*ff*.
Dominance, 57, 72, 177*ff*.
Dominant factors, complementary action of, 171
Double cross, 223
Double fertilization, 203.
Draba, 155
Drosophila, 179, 188, 198, 208
   melanogaster, 61, 111, 184
   sterility in, 112
Düsing, 106

Echinodermata, 22
Echinoderms, 21
Edwards, 235
Ellis, 233
Emerson, 183
Endogamy, 238
Endosperm fertilization, 153
England, 257*ff*., 260*f*.
Englishman, 250, 262
Epilepsy, 231, 241
Eschscholtzia, 117
Europe, 247, 251*f*., 257, 261
   leaders of, 257
European culture, 256
Evolution, 13
   inbreeding and outbreeding in, 195*ff*.
Exogamy, 13, 15, 201

Factors, stability of, 76*ff*.
Faraday, 235
Feeble-mindedness, 231
Ferns, 29
Fertilization, diagram to illustrate, 39
   double, 41
   in embryo sac of the lily, 41

INDEX

Fish, 92, 158
Fitzhugh, 235
Flat worms, 21
Focke, 141, 145
Fossil ants in amber of Oligocene period, 77
France, 257f., 260f.
Franklin, 235ff.
Franks, 260
Frazer, 13
Freeman, 157
French, 257
Frenchman, 250
Frequency distribution of corolla length in tobacco, 70
Freud, 13
Fucus, 26, 28
Funaria, 46

Galton, 50, 227, 233, 237
Gamete formation in dihybrid, 59, 63
Gametogenesis, 38, 55
Gametophyte, 29
Gardner, 118
Gärtner, 141, 143, 145, 155
Gaur, 192
Gayal, 192
Genetics, 50
Genotype hypothesis, 166
Gentry, 213
Germany, 257, 260, 262
Germplasm, mixture of, 201
Gernert, 155
Gerschler, 158
Geum, 144
Goat, 212
Goddard, 240, 243
Goliath, 110 (fig. 28), 233
Gonochorism, 22, 33
Grape, 210
Gravatt, 192
Greece, 251
Green Algæ, 26
Guaita, von, 101ff.
Guinea-pig, 188
  growing curves of, 160
Guinea-pig, inbreeding experiments with, 110ff.

Haiti, 253
Hammurabi, code of, 14
Hare-lip, 230
Hawkweed, 22
Hayes, 89, 148, 168, 192
Herbert, 141
Heredity coefficient, 196, 199
Hermaphroditism, 22ff., 26, 30, 33, 201
Hero, 117
Heterosis, 16, 96, 141, 144, 157, 172, 202
  importance of in sex origin, 204ff.
  manifestations of, 150
  selective effect of, 154
Heterozygosis, 121, 138
  degrees of, 93
  similarity of effect of, with environment, 157
Hittites, 261
Homozygosis, 113
Homozygosity, 134
  affected by linkage, 95
  attainment of, 95
Horse, 212
  origin of, 210ff.
Huntington's chorea, 230
Huth, 243
Hybrid vigor, 16, 88, 96, 141
  benefit from, 219
  cause of, 164
Hydatina, 158
  senta, 112f.
Hyde, 112, 158
Hyoscyamus, 155

Iberian, 259
  race, 258
Ichythyosis, 230
Inbreeding, curves of, 84
  effect of, on yield and height of maize, 124ff.
  effect on organisms, 137

Inbreeding, experiments with animals and plants, 100
   index, 81
   intensity of, 85
   mathematical considerations of, 80
   problem, phases of, 81
   reduction in vigor resulting from, 96
   and outbreeding in plant and animal improvement, 210
Infant consultations, 227
Infusoria, 78
Inheritance, Mendelian, 72
Insanity, 231, 241
Insects in Oligocene amber, 197
Ipomea, 115, 117
Ireland, 257*ff*.
Irish, 260
Isopods, 24

Japan, 247
Japanese, 254*f*.
Jennings, 78, 92, 95, 97, 202*f*.
Jew, 261*f*.
   English, 262
   German, 261
   Spanish, 261
Johannsen, 78, 166
Jukes, 237*ff*., 260, 262
Jutes, 259

Keeble, 170, 172
Kempton, 153
Kerner, 34
King, 101*ff*., 105, 107, 160, 188
King Melia Rioter 14th, 85*ff*.
Knight, 114, 141*ff*.
Knight-Darwin law, 33
Knuth, 34
Kölreuter, 141*f*., 144, 154

Lang, 13
Lavatera, 144
Leaming strains, 124
Lecoq, 141

Lee, 235
Leguminosæ, 114
Lethal factors, 179
Linaria, 144
Lincoln, 235*f*.
Lindley, 143
Liverworts, 29, 45
Lobelia, 144
Loeb, 200
Lychnis, 144
Lycium, 144
Luffa, 144

MacLennan, 13
MacNamara, 257
Maize, connate seeds of, 133
   distribution of rows of grain of, 129
   fertility of, 188*ff*.
   growth curves of, 152
   inbred strains and hybrids of, 150 (Fig. 31)
   inbred strains of, 130 (Fig. 29)
   number of nodes of, 150
   segregation of ear row number of, 131, 132
   variety crosses of, 153
Malthus, 247
Malva, 144, 155
Mammals, crossing of, 159*ff*.
Man, inbreeding and outbreeding in, 226
   interfertility of, 246
Marchals, 46
Marriage of near relatives, 100
Matings, brother and sister, 97
   parent and offspring, 97
Mauz, 141
Mecca of politically oppressed, 264
Mechanism of heredity, 50
   of reproduction, 36
Mendel, 50, 118, 144*f*., 165
Mendelian segregation, 88
Mendelism, 51*ff*.
Mendel's laws of inheritance, 55
Merz, 233

INDEX

Mice, 188
Middleton, 78
Miela, 227
Milk depots, 227
Mimicry, 146
Mimulus, 115, 117
Mirabilis, 142
Mœnkhaus, 112, 158
Molluscoids, 21
Molluscs, 21
Mongolian, 255, 257
Monœcism, 33
Morgan, 62, 198
Morphology, comparative, 18
Morrow, 118
Moss, 29, 46, 204
Mule, 142, 219f.
Müller, 34
Muller, 158
Mumford, 214
Myxomycetes, 26

Nam, 237f., 260, 262
Naudin, 144
Nearing, 233
Negro, 252ff.
Nemathelminthes, 22
Nematode, 21
Nemec, 203
Neolithic period, 258
New Zealand, 247
Nicotiana, 103, 139, 142, 144, 148, 155, 157, 192, 196f.
   alata, 191f.
   height of species and crosses, 149
   Langsdorffii, 191
   longiflora, 69
   paniculata, 192
   rustica, 192
   tabacum, 192
Nordic factors, 250
Normans, 260
Norsemen, 258
Nucleus, 36

Oats, 114

Oögenesis, 37
Open door, 265
Ostrich, 210

Papaver, 144
Paramecium, 202
Parasitism, 22
Parthenogenesis, 206
Pasteur, 235, 237
Payne, 235
Pearl, 80f., 83ff., 87
Pearson, 92
Peas, 148
Pellew, 170, 172
Pentstemon, 144
Peredineæ, 27
Petunia, 117, 144
Phaseolus, 139
Phillips, 159
Pisum, 139
Pœllman, 238
Pollen grains, formation of in the lily, 40
Polydactyly, 230
Polypodiaceæ, 32
Preston, 235
Primula, 144
Protandry, 23, 135, 201
Protogyny, 23, 201
Protozoa, 21
Pumpkin, 221
Punnett, 159

Quantitative characters, inheritance of, 66ff.

Races, intermingling of and national stamina, 245ff.
Racial types, 250
Radish, 192
Ranunculaceæ, 32
Raphanus sativus, 192
Rat, 188
   curves showing body weight of, 107f.
   inbreeding experiments with, 101ff., 105

Rat, size of litters of, 109
Reduction of heterozygous individuals and allelomorphic pairs, 90
Remick, 92
Reproduction, among animals and plants, 20
   asexual, 17, 21, 22, 30, 78
   sexual, 17, 20*ff.*, 26, 201, 205
   sexual, origin of, 26*ff.*
Retina, pigmentary degeneration of, 231
Rhopalura, 26
Rice, 114
Ripley, 257
Ritzema-Bos, 101*ff.*, 188
Roberts, 153
Robbins, 92
Roman Empire, 259
Romans, 259
Rome, 251
Rommel, 110
Rosaceæ, 32
Rotifer, 158, 170, 202

Sacculina, 23
Sageret, 141, 144, 156, 192
Sax, 157
Saxons, 258, 260
Scandinavians, 257
Schizophytes, 26
School for mothers, 227
Scotch, 259*f.*
Scotland, 257, 260
Sex, determination of, 43*ff.*
Sexual dimorphism, 26
Self-fertilization as means of obtaining homozygosity, 97
Self-sterility, 25, 33
Sex-linked characters, 47*ff.*
   inheritance of, 48*f.*
Sex, origin of, 201
Sex ratio, 106
Shamel, 119
Sheep, 210*ff.*
Shull, A. F., 112*f.*, 158, 169
Shull, G. H., 118*ff.*, 168*f.*

Silkworms, 158
Spain, 257*f.*
Spaniard, 262
Spermatogenesis, 37
Spermatozoön, entrance of, through membrane of egg, 42
Sphærocarpus, 45
   Donnellii, 45
   texanus, 45
Spirogyra, 27
Sponges, 21*ff.*
Spores, 29
Sporophyte, 29
Squashes, 221
Sterility, 188*ff.*
Stevenson, 233
Stone Age, 258
Strasburger, 45
Sturtevant, 184
Stylonychia pustulata, 78
Swine, 210*f.*, 213
Synapsis, 37
Syndactyly, 230

Talmud, Hebraic, 14
Tapeworm, 23*f.*
Tobacco, 114, 148, 156
   corolla length of, 70
Tomatoes, 114, 148, 221
Toyama, 158
Trochelminthes, 21*f.*, 202
Tropæolum, 144
Tunicates, 23
Turanian, 259; race, 258
Turbellarians, 23
Tuttle, 235

Ulothrix, 26, 28
Unit of heredity, 77
United Kingdom, 257
United States, 253*ff.*, 260, 262, 264
Ustilago maydis, 118

Venable, 235
Verbascum, 142, 144

Wales, 261
Weismann, 101*ff*., 188, 201, 206
Wentworth, 92, 112
Westermarck, 238, 243
Wheat, 114, 197, 210
Wheeler, 77
Whitney, 112, 158
Wiegmann, 141, 144, 154
William the Conqueror, 260
Woods, 233

Wooley, 235
Wright, 110*f*., 160*f*., 188

Xeroderma pigmentosum, 231

Yak, 192
Yellow mouse, 179
Youatt, 220

Zero, 237, 239

www.ingramcontent.com/pod-product-compliance
Lightning Source LLC
Chambersburg PA
CBHW081142180526
45170CB00006B/1888